Human Factors Considerations in the Design and Evaluation of Electronic Flight Bags (EFBs)

Version 1: Basic Functions

Prepared By:

Divya C. Chandra
USDOT - Volpe Center
Operator Performance and Safety Analysis
Flight Deck Human Factors

Susan J. Mangold
Battelle Memorial Institute

Federal Aviation Administration
AAR-100
Washington, D.C

Final Revision
DOT-VNTSC-FAA-00-22
28 September 2000

For:
Dr. Tom McCloy
Office of the Chief Scientific and Technical Advisor for Human Factors

Version 1 (9/28/00)

Preface

This report contains human factors considerations for the design and evaluation of "Electronic Flight Bags" (EFBs). An EFB is an electronic information management device for use by pilots in performing flight management tasks. It typically consists of a screen and controls in a self-contained unit that is relatively small, weighing only a few pounds. EFBs could support a variety of functions, including electronic documentation, electronic checklists, and flight performance calculations. These EFB functions are addressed in this report. More advanced EFB functions will be addressed in Version 2 of this document, due out in 2001.

This work was conducted at the Volpe National Transportation Systems Center (Volpe Center) under the sponsorship of the Federal Aviation Administration's Office of the Chief Scientific and Technical Advisor for Human Factors. Tom McCloy served as the FAA program manager.

The authors would especially like to thank Bill LeRoy (Chairman) and the other members of the Air Transportation Association Digital Data Working Group for reviewing the document and providing valuable feedback.

This document was prepared by the Operator Performance and Safety Analysis Division of the Office of Research and Analysis at the Volpe Center. It was completed under the Division's Flight Deck Technology Human Factors program.

Version 1 (9/28/00)

Table of Contents

Preface i
Table of Contents ii
Executive Summary v
1   Introduction 1
   1.1   Background 1
   1.2   Overview 1
   1.3   How to Read This Document 3
2   System Considerations 4
   2.1   General 5
      2.1.1   Stowage Area for Portable Units 5
      2.1.2   Design and Placement of Structural Cradle 6
      2.1.3   Crew Knowledge of Revision Dates 7
      2.1.4   Legibility of Text—Lighting Issues 8
      2.1.5   Using EFBs During Critical Phases of Flight 9
      2.1.6   Use and Compatibility of EFB with Other Flight Deck Systems 10
      2.1.7   Alerts and Reminders 11
      2.1.8   Updating EFB Software and/or Databases 12
      2.1.9   Graphical Icons 13
      2.1.10   Supplemental Audio 14
   2.2   Training/Procedures 15
      2.2.1   Part 121 and Part 135 Operations EFB Policy 15
      2.2.2   EFB Documentation 16
      2.2.3   User Feedback 17
      2.2.4   EFB Training for Part 91 Operators 18
      2.2.5   Initial EFB Training for Part 121 and Part 135 Operators 19
      2.2.6   Evaluation Process for Part 121 and Part 135 Operators..................................... 20
      2.2.7   Fidelity of EFB Training Device 21
      2.2.8   Ensuring Data Integrity 22
      2.2.9   Use of Hand-Held EFBs 23
   2.3   Equipment 24
      2.3.1   Input Mechanisms 24
      2.3.2   System Error Messages 25
      2.3.3   Compatibility Across Applications on the EFB and Use of Style Guides 26
      2.3.4   Multi-Tasking 27
      2.3.5   Responsiveness of Application 28
      2.3.6   Soft Keys 29
      2.3.7   Anchor Locations 30
      2.3.8   Legibility of Text—Character Issues 31
      2.3.9   Legibility of Text—Typeface Size and Width 32
      2.3.10   Legibility of Text—Spacing for Readability 33

3 Electronic Documentation 34
3.1 Background 34
3.1.1 Type of Documents Addressed 34
3.1.2 Features of Electronic Documents 34
3.2 General 37
3.2.1 Consistency of Logical Structure Between Paper and Electronic Documents 37
3.2.2 Consistency of Electronic Document User Interface 38
3.2.3 Training Needs 39
3.2.4 Speed of Loading Data 40
3.3 Layout/Appearance 41
3.3.1 Visual Structure 41
3.3.2 Minimum Display Area 42
3.3.3 Off-Screen Text 43
3.3.4 Active Regions 44
3.3.5 Display of High Priority Information 45
3.3.6 Figures 46
3.3.7 Tables 47
3.4 Navigation and Searching 48
3.4.1 Moving to Specific Locations 48
3.4.2 Managing Multiple Open Documents 49
3.4.3 Searches 50
3.5 Options 51
3.5.1 Links to Related Material 51
3.5.2 Display Customization 52
3.5.3 Printing 53
3.5.4 Animation 54
3.5.5 Making Notes 55
3.5.6 Decision Aid/Automatic call-up of Data 56

4 Electronic Checklists 57
4.1 Background 57
4.2 Call-up/Access 59
4.2.1 Checklist Scope 59
4.2.2 Accessing Normal Checklists 60
4.2.3 Accessing Non-normal Checklists 61
4.2.4 Open Checklists 62
4.2.5 Multiple Open Checklists 63
4.2.6 Managing Multiple Checklists 64
4.2.7 Managing Multiple Non-Normal Conditions 65
4.2.8 Putting Away the Checklist 66
4.3 Checklist Actions 67
4.3.1 Indicating the Active Checklist Item 67
4.3.2 Moving Between Items Within a Checklist 68
4.3.3 Specifying Completion of Item 69

4.3.4 Closing a Checklist	70
4.3.5 Undoing An Item Status Change	71
4.3.6 Displaying Item Status	72
4.3.7 Returning to Deferred Items	73
4.3.8 Integrating Non-Normal Items into Subsequent Checklists	74
4.3.9 Lengthy Checklists	75
4.3.10 Confirming Completion of Checklist	76
4.4 Optional Features	77
4.4.1 Links Between Checklist Items and Related Information	77
4.4.2 Links to Special Information for Ongoing Non-Normal Conditions	78
4.4.3 Links to Calculated Values	79
4.4.4 Task Reminders	80
4.4.5 Checklist Branching	81
5 Flight Performance Calculations	82
5.1.1 Aircraft Performance Documentation	82
5.1.2 Data-entry Screening and Error Messages	83
5.1.3 Support Information for Data Entry	84
5.1.4 When and How to Do Performance Calculations	85
5.1.5 Default Values	86
Appendix A: Related Literature	87
Appendix B: Acronyms	89

Executive Summary

There is currently great interest in developing electronic information management devices for use by pilots in performing flight management tasks. These devices are sometimes referred to as "Electronic Flight Bags" (EFBs). EFBs typically consists of a screen and controls in a self-contained unit that is relatively small, weighing only a few pounds at most. They were originally seen as a repository for electronic documents such as checklists, operating manuals, and navigation publications. In the future, many airlines envision that EFBs may become multi-function devices supporting an array of applications beyond those of a traditional flight bag, from electronic messaging to display of live weather.

The Federal Aviation Administration (FAA) is charged with approval of EFBs for installation and operational use in aircraft. The approval process will be a multi-dimensional effort requiring an understanding of how the device functions and is used by crews, how the device interacts with other flight deck equipment, and training and operating procedures. This document is intended to facilitate the development of FAA advisory material for evaluation and operational approval of EFBs. The document will also be accessible to users and avionics manufacturers for use in system design and development. Note that the regulatory application of this information is the responsibility of the appropriate government agencies. In the United States, the Federal Aviation Administration plans to publish an Advisory Circular for Electronic Flight Bag equipment that is intended to reference this document.

Version 1 of this human factors document covers system considerations and three EFB functions in detail. The initial set of functions discussed in Version 1 are electronic documentation, electronic checklists, and flight performance calculations. These functions were chosen because they are the most mature applications to date. Advanced EFB functions, such as display of electronic approach plates, will be addressed in a Version 2 of this document, to be released in 2001.

1 Introduction

This document contains human factors considerations for the design and evaluation of the Electronic Flight Bag (EFB). The regulatory application of this information is the responsibility of the appropriate government agencies. In the United States, the Federal Aviation Administration (FAA) plans to publish an Advisory Circular for Electronic Flight Bag equipment that is intended to reference this document.

Chapter 1contains background information on existing draft advisory material from the Federal Aviation Administration and an overview of the structure of this document. There are separate chapters for system considerations, which are independent of the function(s) supported by the EFB, and three specific EFB functions. The three functions addressed in this document are, electronic documentation, electronic checklists, and flight performance calculations. These functions were chosen because they are the most mature applications to date. Several airlines are in the process of converting manuals into electronic form, the first step towards placing the documents on a portable flight deck device. Electronic checklists are already available on some newer aircraft models. The algorithms for computing flight performance are well understood, and have been implemented for use on standard personal computers. Other functions will be addressed in Version 2 of this document, due out in 2001.

1.1 Background

To date, the FAA has released several preliminary documents on the approval of EFBs (see list of FAA documents in Appendix A (Related Literature). The most recent is the draft advisory circular (AC) from Flight Standards (AFS-400), AC20/120-EFB (July 2000). This draft advisory circular incorporates information from a September 1999 memorandum from Aircraft Certification. It contains a flow chart, reproduced below in Figure 1-1, which can be used to determine when Aircraft Certification needs to be involved in the approval process.

There are three types of EFBs mentioned in draft AC20/120-EFB. All three types require the involvement of Flight Standards for the operational approval of the devices. The first type is a self-contained device that does not interface electronically or mechanically with the aircraft; these are considered to be portable electronic devices subject only to Flight Standards involvement in the approval process. The next type of EFB is one that is mechanically connected to the aircraft via a structural cradle. This type of EFB is subject to involvement by both Flight Standards and to a lesser extent, Aircraft Certification. The last type of EFB has some electrical interface with the aircraft, either for power or data communications. As shown in Figure 1-1, if any sort of data communications interface is present (either via the aircraft avionics data bus, or via a proprietary EFB communications network), there is a significant type design impact, requiring a much higher level of involvement of Aircraft Certification, as well as involvement of Flight Standards for the operational approval, or approval of the equipment use in an operational environment.

From a human factors perspective, these distinctions are important in that they denote *two* major levels of capability; devices that communicate with the aircraft systems allow for much more integrated functionality between the EFB and other avionics than those that do not. These EFBs could sense the status of aircraft systems and automatically bring up information, if necessary, to address an abnormal situation. Throughout this document, we distinguish issues relevant only to EFBs that are integrated with other aircraft systems from issues relevant to EFBs that do not communicate with aircraft subsystems.

1.2 Overview

This document contains four sections with guidelines on specific topics for the FAA and avionics manufacturers (Sections 2 through 5). Section 2 contains system considerations while Sections 3, 4, and 5 cover the three basic applications, electronic documentation, electronic checklists, and flight performance calculations, in order. Appendix A contains a list of references, and Appendix B contains a list of acronyms.

Version 1 (9/28/00)

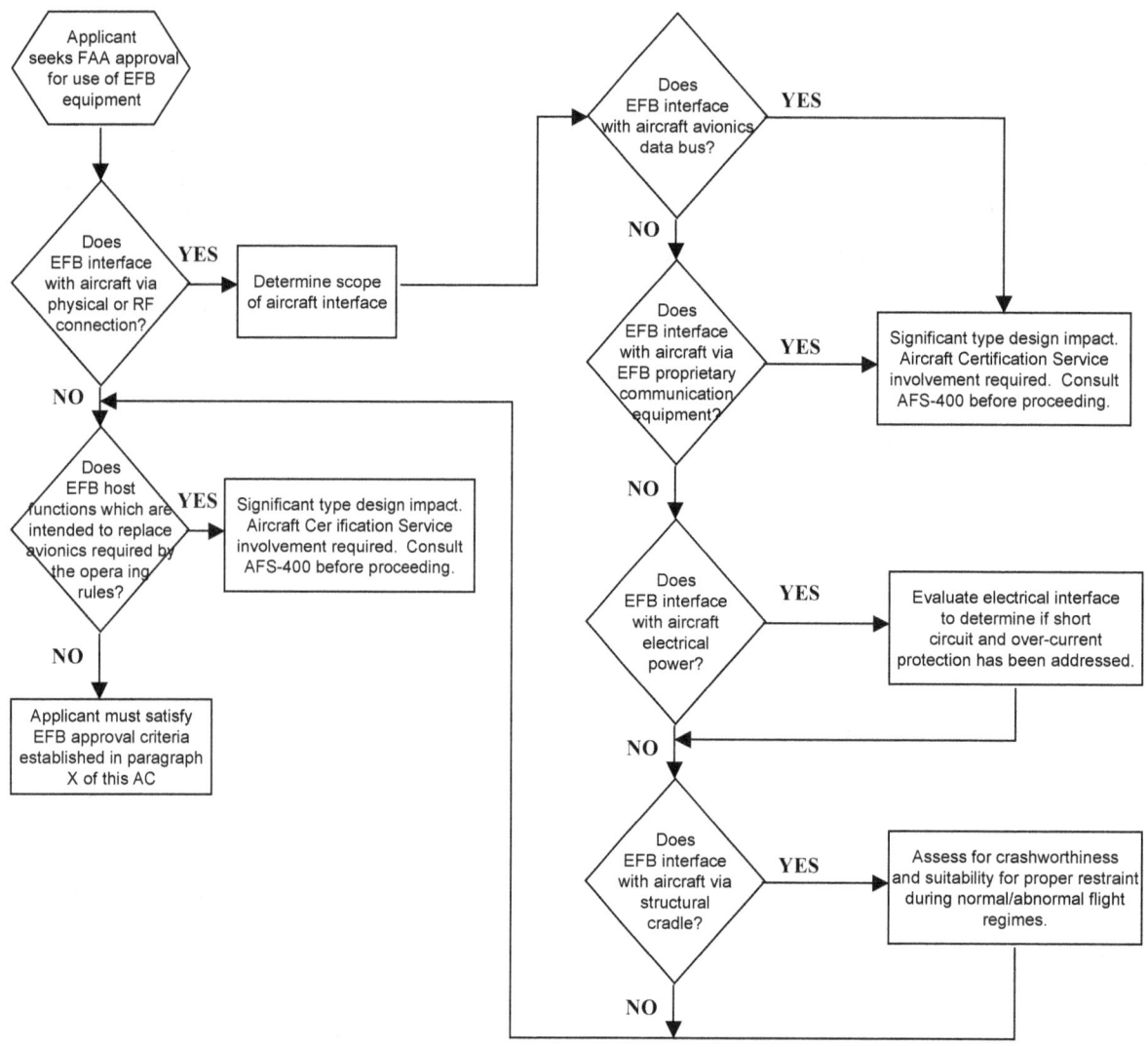

Figure 1-1 Draft flow chart of how to decide when Aircraft Certification Service will be involved in approval of EFBs.

Section 2 (System Considerations) contains several individual "considerations," the format of which is described below. Section 3 (Electronic Documents) begins with background about the different types of document capabilities. Readers who are familiar with these concepts and terms could skip this section and move directly to the considerations that follow. Section 4 (Electronic Checklists) also begins with background material on electronic checklists, and a description of terms. Again, readers who are familiar with these terms can move directly to the considerations that follow.

Each consideration may have one *or more* boxed summary statements at the top of the page. Each of the statements is preceded by a descriptive label, such as "Equipment Requirement", or "Installation Recommendation." The purpose of these labels is to identify (a) what aspect of FAA approval the statement concerns , and (b) the type of information in the statement.

The three descriptive labels used are: "Equipment," "Installation," or "Training/Procedures." These categories are separated out to assist readers who are coming from a specific one of these perspectives to browse the document. Equipment guidelines are "box level" design items which can typically be assessed in a bench test; i.e., these items are testable outside the context of the aircraft. Installation issues are those that must be tested within the context of the aircraft flight deck. Because EFBs may be portable, some issues may be both "Equipment" and "Installation" issues. Training/Procedures issues are, of course, related to training and procedures rather than the design or installation of the equipment itself.

Version 1 (9/28/00)

The information within a boxed summary statement in a consideration can be one of four types: requirements, recommendations, good practices, and issues. Each type of information is designated by a different style of text, as shown in Table 1.2.1. As indicated by the labels, requirements are mandatory and recommendations denote a preferred method or mechanism. Good practices are suggestions based on generally well established principles in use. "Issues" differ from the other types in that there is less specific information to deal with, but the item should be addressed.

Each consideration is covered in one page. To browse the document, read just the summary statements in each of the boxes. For more information about that issue, read the rest of the page, which has more detailed information about the consideration in one or more of the sections below the summary statements. These sections are labeled: Problem Statement, Example(s), and Evaluation Questions. The Problem Statement describes the problem that the summary statements apply to, including the potential impact if the problem is not addressed. The Example(s) section contains examples of the potential problem, and possible solutions. The Evaluation Questions section lists open ended questions that an FAA inspector should consider when determining whether the problem has been adequately addressed. The Evaluation Questions do not provide detailed guidance on performance assessment, but they do point out areas for evaluation.

Table 1-1 Elements of each Consideration

Requirements are shaded and boxed with a bold outline. These are the items the authors feel are mandatory. However, it should be noted that this is not a regulatory document and that any application of these requirements is the responsibility of the appropriate regulatory agency (such as the FAA in the United States).

Recommendations are boxed within a bold outline. These are preferred methods or mechanisms.

Good practices are boxed within a thin outline. These are suggestions based on industry practices.

Issues are boxed within a double line. These point out design tradeoffs and other related factors in an open-ended statement.

1.3 How to Read This Document

This document is designed to be read by both FAA personnel (from various offices) and avionics manufacturers. From the manufacturer's perspective, each consideration should be read completely, starting with the summary statements (requirements, recommendations, etc.). The rest of the material (Problem Statement, Examples and Evaluation Questions) should be helpful in illustrating the spirit of the guidelines.

Because the FAA readers are themselves a diverse audience, sections are labeled for ease of browsing. Aircraft Certification personnel should focus more on Equipment and Installation Requirements. Flight Standards personnel should find topics of relevance throughout, with particular emphasis on Equipment and Training/Procedures requirements. Principal Operations Inspectors (POIs) and Principal Avionics Inspectors (PAIs) may find the Evaluation Questions particularly important.

2 System Considerations

System considerations apply to any EFB, regardless of the function, or functions, performed by that EFB. These considerations are divided into three categories: General, Training/Procedures, and Equipment. Section 2.1 contains general considerations, which include installation issues, and issues that cut across the installation, training/procedures, and equipment audiences. Readers who are interested only in installation topics need only review Section 2.1. Readers interested in training/procedures issues should review both Section 2.1 and 2.2. Readers coming from an equipment perspective should review Sections 2.1 and 2.3.

2.1 General

2.1.1 Stowage Area for Portable Units

Installation Requirement(s)

> A stowage area with a securing mechanism for the EFB is required for storage of portable units when they are not in use.
>
> Note: If the EFB is designed to be held in a structural cradle, the cradle may satisfy the requirement for a stowage area.
>
> Note: For EFBs that are not used as the only means of performing any tasks that are critical for the safety of flight, this requirement may be downgraded to a recommendation.

Training/Procedures Recommendation(s)

> Crews should routinely store EFBs that are not in use.

Problem Statement

Flight deck real-estate (not just display space) is extremely limited. Every device routinely used in the flight deck must have a designated place, both when in and out of use. Portable devices that do not have designated locations can be a hazard because they may create confusion when crews attempt to locate, orient, and use them. They may also be a hazard in the case of strong accelerations, such as those in takeoff, landing, and turbulence. For example, a unsecured EFB could fall and jam rudder pedals or limit aft yoke travel. Unsecured units could also cause physical injury to the crew under these conditions.

Example(s)

Unsecured units may move unexpectedly during significant accelerations. For example, a unit left on an unused seat may fall off the seat during turbulence. The next time the pilot attempts to use the device, finding the unit will cause pilot distraction at the least.

During takeoff and landing, the EFB may need to be stowed in order to prevent injuries to the crew in case of sudden aircraft accelerations, similar to the requirement for stowing tray tables for passengers.

Evaluation Questions

- Is there a stowage area for the EFB? When the EFB is not stowed, is the securing mechanism in the stowage area unobtrusive?
- When the device is stowed, does the combination of it and the securing mechanism intrude into any other flight deck spaces, causing either visual or physical obstruction of important flight controls/displays?
- Does movement of the EFB to and from a stowage area require substantial effort, or substantially limit access to flight displays and controls? Is the securing mechanism simple to operate?

Version 1 (9/28/00)

2.1.2 Design and Placement of Structural Cradle

Installation Requirement(s)

> The structural cradle must not be mounted such that it substantially obstructs visual or physical access to critical flight controls and/or displays. Note: A structural cradle is a piece of hardware that is designed to hold the EFB and is physically attached to the aircraft.
>
> If the EFB can be used while held in a structural cradle, then the cradle must also allow for some adjustment of the EFB orientation in order to allow users to customize their viewing angle.
>
> If the EFB can be used while held in a structural cradle and it is used during critical phases of flight (e.g., takeoff, landing, or emergencies), then the cradle must be mounted such that the pilot can use the device without turning more than 90° to either side..

Installation Recommendation(s)

> If the EFB is used while held in a structural cradle, then the design of the cradle should allow the user easy access to all the EFB controls and a clear view of the EFB display. Also, the cradle should be mounted such that the pilot does not have to turn him- or herself significantly to use the device, particularly if it is used during critical phases of flight.

Training/Procedures Requirements(s)

> Crews must know how and when to adjust the cradle position.
>
> Procedures must ensure that pilots have good access to all flight/controls and displays, even those that are partially obstructed by the EFB and its cradle when necessary.

Problem Statement

New devices that are added into older flight decks could obstruct access to or use of other critical equipment.

A structural cradle can help assure that the device is positioned appropriately for use in a flight deck, but it must also be flexible in order to accommodate viewers of different heights, who will see it from different design-eye viewpoints.

During critical phases of flight, it is important that the pilot be able to continue his/her scan out the window. Therefore, the pilot must not be required be use an EFB that is located behind him/her.

Example(s)

A short pilot and a tall pilot will need to view the EFB from different positions. Both of these pilots, however, should not place the EFB in the viewing path obstructing critical flight controls/displays.

Evaluation Question(s)

- Does the structural cradle substantially obstruct visual or physical access to critical flight controls and/or displays? Which controls/displays are affected, and how critical are they during the different phases of flight in which the EFB will be used? Can the cradle be adjusted so that these controls/displays are more accessible if necessary?
- Do crews know how to adjust the EFB orientation be adjusted? Is there adequate room to manipulate the device controls and view its display?
- Is the EFB positioned in front of the pilot during critical phases of flight?

2.1.3 Crew Knowledge of Revision Dates

Training/Procedures Requirement

> A procedure must be in place for crews to confirm that the latest revision of EFB databases and software are installed in their units for each flight. Pilots should know how to obtain the latest software and/or databases if they find that the their unit is out of date.

Training/Procedures Recommendations

> The procedure for checking revision dates should be consistent across the different applications available on the EFB, and consistent with airline/operator standard operating procedures.
>
> Change information explaining the updates made in the latest revision should be provided to the crews.

Equipment Requirement

> The EFB must provide the latest revision information to the crew upon request.

Problem Statement

The crew is ultimately responsible for the safety of the flight; therefore, they are responsible for ensuring that they use the most up-to-date information available in conducting the flight. A procedure needs to be in place to make sure that this task is as simple and error-proof as possible for the crew.

Since EFB software and databases may change at different times for the different applications, making sure that the most up-to-date information is in place can become even more complex if the procedure for checking revision dates differs between applications. Therefore, it is best to keep the procedure consistent across the different EFB functions. Ideally, revision information for all EFB functions would be available in one place.

Example(s)

Similar to the requirement that crews ensure they are using the latest revision of an approach plate, crews must ensure that they are using the latest version of EFB software and data. Procedures that apply to checking dates for approach plates could be adapted for use with the EFB applications.

Evaluation Questions

- What is the procedure for ensuring that data stored in the EFB is up to date? Is the procedure for checking the EFB data revision dates consistent with other standard operating procedures?
- Can the crew request revision information from the EFB? Is the revision information presented clearly?
- Are procedures in place so pilots know what to do with an out of date database?

2.1.4 Legibility of Text—Lighting Issues

Installation Requirement

> Text on the EFB must be legible by the typical user under the variety of lighting conditions expected in a flight deck.

Installation and Equipment Requirement(s)

> Users must be able to adjust the screen brightness of an EFB independently of the brightness of other displays in the flight deck.

Equipment Requirement

> If the EFB will be used away from the flight deck (e.g., outdoors, or in a home/office setting), text on the EFB must be legible by the typical user under the expected lighting conditions for use.

Equipment Recommendation

> Screen brightness should be adjustable in fine increments or a continuous rather than discrete manner.

Problem Statement

Ambient lighting conditions may vary from very dark during a night flight to very bright in sunny conditions. The EFB display must be usable under all of these lighting conditions. This will require at least some ability to adjust the screen brightness. In addition to screen brightness, size of the text, font style, and viewing distance, and off-angle viewing will all affect text legibility. Because the EFB may be portable, legibility should be checked both inside and outside the flight deck.

Screens or text that are not legible will cause pilot distraction at the least (as the pilot attempts to position the display for better legibility) and potentially more harmful consequences if critical information is misread, or not read at all. It is particularly important to assure that all colors on the screen are legible under the given conditions. There may be special considerations for EFBs to be used with other devices, such as head-up displays or night vision goggles, where additional lighting issues need to be considered.

Example(s)

Different font styles may be used at the same time on an EFB. The different fonts may indicate something about that information. For example, one font style may denote the active checklist item (e.g., white, large font) and another font style could represent an completed item (e.g., green, small font). Because the font styles can encode important information, all of the font styles on the screen must be legible and easily discriminated from one another without any screen adjustments.

Evaluation Question(s)

- Can the EFB screen be read under a variety of typical flight-deck lighting conditions? Can the EFB screen be read outside the flight deck?
- Can the user adjust the screen brightness and contrast (if applicable)?
- If colors are used, are they all legible and discriminable under the different possible lighting conditions? Can colors be viewed and discriminated by all users, even those who view the screen off-angle?

2.1.5 Using EFBs During Critical Phases of Flight

Training/Procedures, Equipment, and Installation Issue

> If the EFB is to be used under critical phases of flight, such as take off and landing, its use under those conditions should be carefully evaluated. Procedures should be designed to mitigate any additional workload of using an EFB. EFB software should be designed to minimize workload as well. The installation of the EFB should also be designed to minimize any possible additional workload incurred by the need to locate and orient the display.

Training/Procedures Recommendation

> Complex, multi-step data entry tasks should be avoided during takeoff, landing, and other critical phases of flight.

Problem Statement

Using an EFB requires effort. There may be effort involved in locating and orienting the display for use and there is effort in looking at the display, processing the information, and making any necessary entries. Data entry can produce particularly long head-down times and high workload. Visual scanning of the EFB (without data entry) does not require as much effort, but it is still an additional task for the pilot. The additional workload required to use an EFB may distract the pilot from higher priority time-critical tasks during critical phases of flight.

When evaluating the workload incurred by use of an EFB, the evaluator should consider factors such as the time it takes to complete the task using the EFB, as compared with doing the task without the EFB. The location and accessibility of the EFB display and controls, and the amount of automation, and the usability of the EFB software will all affect the time it takes to complete a task using the EFB. The evaluator should also consider whether users would be more likely to make errors during high workload conditions, whether it is easy to recover from errors, and whether users are likely to become distracted from other flight deck tasks while resolving EFB problems.

Example(s)

During high workload situations, such as takeoff and landing, entering data on the EFB may distract the crew from essential functions, such as visual scanning for air traffic out the window, or scanning of aircraft instruments. Data entry tasks should be avoided during these phases of flight. If data entry is required, it should be limited to a single key press. For example, to indicate that the "climb out" checklist has been completed, the pilot may enter a yes/no response to an EFB inquiry.

If, however, the EFB is used as a display of real-time information useful during landing (e.g., if the EFB displays nearby traffic during landing), and only requires occasional scanning that the pilot can incorporate into his/her task schedule, the additional workload may be acceptable.

An EFB that has more built-in automation may also be more acceptable for use under high workload conditions. For example, if some items in an emergency checklist are completed through closed-loop sensors, the pilot's workload may not be impacted negatively by using the EFB as compared with the paper checklist.

Evaluation Question(s)

- What training and/or procedures are in place to mitigate any additional workload of using the EFB under high workload conditions?
- Is the EFB installed for easy access during critical flight phases?
- Is the software function designed to minimize any additional workload of using the EFB?

2.1.6 Use and Compatibility of EFB with Other Flight Deck Systems

Training/Procedures Requirement(s)

> The crew must know what flight deck automation system (e.g., EICAS, FMS, or EFB) to use for a given purpose, especially when there are similar functions across the systems. Policy must address which information/data will be used, when information/data is provided by both aircraft and EFB systems (e.g., aircraft performance).
>
> If the EFB can generate information that can also be generated by existing cockpit automation (e.g., a flight performance variable), there must be clear procedures establishing which information to use and under what conditions the backup source of information should be used.

Equipment Issue

> The user interface of the EFB should match the flight deck design philosophy. For example, terms for specific types of data or the format in which latitude/longitude is entered should be the same across systems. Data entry methods should be as consistent as possible between the EFB and other flight deck systems. Color coding philosophies and symbology should also be consistent across systems.

Problem Statement

Whether or not there is any communication between aircraft systems and the EFB, from the perspective of a crew member, the EFB is just another tool for him/her to use. If there are inconsistencies or redundancies in the information provided by the different automation systems ("tools"), there will be confusion and increased potential for errors.

Therefore, regardless of whether there is a data connection between the flight deck systems and the EFB, information consistency/redundancy must be considered when integrating an EFB into an aircraft with other advanced systems, such as a Flight Management System (FMS) or Engine Indication and Crew Alerting System (EICAS).

Example(s)

EFBs may support electronic checklists. On some Airbus aircraft, electronic checklists for emergencies are built into the Electronic Centralised Aircraft Monitor (ECAM), and paper checklists are provided as well. Procedures must be established to ensure that crews know which of these checklists should be used in an emergency.

In another example, it is possible that the EFB and FMC could both compute flight plans. Crews must know which flight plan is to be used for the flight, and they must ensure, if appropriate, that this is the plan in use by automation on the aircraft (e.g., the FMC and/or auto-pilot).

Evaluation Question(s)

- What are the procedures for establishing which source of information is primary and under what conditions the backup source of information must be used?
- Does the user interface of the EFB match the philosophy of the flight deck designers? Are the terms consistent? Are the color codes and symbols consistent? Are the data entry methods consistent?

2.1.7 Alerts and Reminders

Equipment Requirement

> EFB alerts and reminders must meet the requirements in FAR 23.1322, 25.1322, 27.1322 or 29.1322 depending on the type of aircraft/rotorcraft.

Equipment Recommendation(s)

> EFB alerts and reminders should be consistent with the FAA standards for electronic displays in AC 25-11 and FAR 23.1311a. International recommended guidance can be found in publications of the Joint Aviation Authorities (JAA) including AMJ 25-11 (Advisory material Joint) and AMJ 25.1322 on warnings and cautions. Additional recommendations are listed in the January 1996 FAA Human Factors Design Guide for Acquisition of Commercial-off-the-shelf Subsystems, Non-Developmental Items, and Developmental Systems—Final Report and Guide (Sections 7 and 8).
>
> Non-critical alerts should be inhibited during critical phases of flight.

Installation and Equipment Issue

> EFB alerts and reminders should be integrated with (or compatible with) presentation of other flight deck system alerts. Designers should consider where the alerts are displayed, particularly for EFBs that communicate with flight deck systems. The alerts could be displayed on the EFB itself, or on a general flight deck alert/reminder display. If some alerts/reminders are shown in one place, and others are shown in another place, there should be a clear rationale, from the crew's perspective, as to what types of alerts/reminders are shown in each location.
>
> Audio alerts may also be useful. Any use of audio should be assessed in terms of possible confusion with other audio alerts.

Problem Statement

Because EFBs may be integrated with flight deck systems, and/or present information that is critical to flight safety, it is important to integrate the EFB alerts/reminders into the overall flight deck alert/reminder philosophy. Some of the factors to consider in the display of alerts is how time-critical they are and their level of priority within the overall context of conducting the flight.

Because the flight deck environment places tremendous demands on the pilot visual system, audio alerts are often used. Audio is already used on the flight deck by a number of systems. Ten or more unique audio sounds, together with dozens of vocal warnings, are not uncommon on advanced automation aircraft. An audio warning system as part of an EFB should be assessed in terms of possible confusion with other systems, ease of control, and training requirements.

Example(s)

Messages that are time critical, although they may be generated by an EFB, should be displayed on a general alerting system.

Messages specific to the EFB application (e.g., "value out of bounds") should be displayed on the EFB.

It may also be appropriate to display messages with different amounts of detail both on the EFB and on the general alerting display. For example, the general alert display could show generic text such as "EFB Message" and the specific message would be shown on the EFB.

See the Terrain Awareness and Warning System (TAWS) TSO C151a for an example alert prioritization scheme.

Evaluation Question(s)

- Do the EFB alerts and reminders meet the requirements in the appropriate FARs?
- What is the philosophy for alerts/reminders? If implemented, are audio alerts used effectively?
- Are colors used appropriately (e.g., red for warnings, as noted in AC 25-11 and FARs 25.1322, 23.1322, 27.1322 and 29.1322))?

2.1.8 Updating EFB Software and/or Databases

Training/Procedures Recommendation(s)

> Procedures should be developed for handling EFB software and database upgrades and customization.

Equipment Issue(s)

> Manufacturers should have a plan for how to handle modifications to the EFB after its initial purchase and installation. This plan should be understood by the customer.

Equipment Good Practice(s)

> Air carriers often customize checklists. If the EFB supports electronic checklists, these should be customizable by the air carrier, rather than requiring modification by the manufacturer.

Problem Statement

Customers expect that EFB software and databases will be upgradable and customizable. It is important that there is clear communication between the manufacturer and the customer about how upgrades and customizations can be performed. The responsibility for performing the EFB modifications should be clearly assigned.

Users should not be required to *remember* what information is current and what is not. Also, paper bulletins should not be the only way that users are informed of changes to EFB software or databases.

Example(s)

Air carriers often customize checklists. If the EFB supports electronic checklists, these could be customizable by the air carrier, rather than requiring modification by the manufacturer.

Evaluation Questions

- Are there procedures establishing who will upgrade and customize the EFB software?

2.1.9 Graphical Icons

Training/Procedures Recommendation(s)

> Users should receive training on the meaning of graphical icons used on the EFB.

Equipment Recommendation(s)

> Users should be able to access text help information to explain meaning of graphical icons.

Equipment Good Practice

> Using the same graphical icons as other flight deck systems can reduce training requirements and help to prevent error, especially in critical phases of flight.

Problem Statement

Graphical icons could be used to access commands from graphical menus/toolbars, or they may represent files and other system objects. Icons can be especially useful for expert users. However, they are typically small and of limited graphical resolution, so the actual object or command they represent may not be intuitively clear to untrained users. Even trained users may forget the meaning of an icon.

Because of their potentially critical role, it is important that users are trained on their interpretation and it is important that the actual graphical image of the icon be a redundant, not sole, means of representing an object or command. Text information about that object or command should also be available.

Example(s)

Achieving familiarity with all EFB icons should be an important goal of initial EFB training. In the event that the user forgets an icon's meaning, the EFB should provide an easy means for finding that information. One approach is to provide a textual label for the icon if the cursor lingers on that icon. Another approach is to provide a "cheat sheet" list of all icons and their meaning that is available in a centralized location. Using icons that are commonly used by other flight deck equipment can reduce the likelihood of a user forgetting an icon's meaning.

Evaluation Question(s)

- Does the initial EFB training adequately address icon meanings?
- Does the EFB provide easy access to help information that explains each icon's meaning?
- Does the EFB use icons in a way that is consistent with other flight deck systems?

2.1.10 Supplemental Audio

Equipment Recommendation(s)

> Note: Supplemental audio is defined as audio that is not associated with alerts and warnings. Supplemental audio could be verbal or non-verbal.
>
> Users should be able to control the volume of supplemental audio. They should also be able to turn off the supplemental audio on an EFB if desired.
>
> Objects that have associated supplemental audio should be coded such that the user knows of the associated audio before it is activated.
>
> Supplemental audio that is audio alone (i.e., without any visual image) should have a text description available so that the user can anticipate the content of that audio clip.
>
> Users should be able to stop the supplemental audio at any time while it is in progress.
>
> Overuse of supplemental audio should be avoided.

Equipment and Training/Procedures Issue

> Supplemental audio may be useful for enhancing animation segments in a multimedia document. It could also be used for training purposes, especially if the sound is of high quality. However, operators should consider whether their policy should be to limit the use of supplemental audio in flight because the additional audio may interfere with higher priority audio information (e.g., radio communications). Also, users may need training in how to use and control any supplemental audio functions.

Problem Statement

Supplemental audio is an optional advanced EFB feature. It has the potential to significantly modernize the look and feel of the EFB. However, because it is an optional features, users must have complete control over when (and whether) the audio is activated, and its volume. Because the flight deck has many other sources of higher priority auditory information, use of supplemental audio in flight may need to be limited. The utility of supplemental audio may be highest for ground-based training purposes.

Example(s)

Video clips of training presentations, complete with supplemental audio, could be stored on the EFB. These could be accessed through a help facility.

Supplemental audio such as background music could be distracting and useless in that it does not convey any additional information to the user.

Evaluation Question(s)

- If supplemental audio is implemented, does the user have control over when, and whether, the audio is activated?
- Is supplemental audio overused?

2.2 Training/Procedures

2.2.1 Part 121 and Part 135 Operations EFB Policy

Training/Procedures Requirement(s)

> Part 121 and Part 135 operators must have a policy that defines how the crew is expected to utilize EFB functionality. The policy must adequately address the specific usage of each EFB function.
>
> The policy must be provided in written form to flight crews, maintenance staff, dispatchers, and other employees whose responsibilities overlap with the functionality supported by the EFB.
>
> Existing policies that could be affected by the introduction of the EFB into line operations must be reviewed to assess whether any modifications to these policies is required.

Training/Procedures Recommendation(s)

> The EFB policy should specifically address use of the EFB and its specific applications under all flight conditions.

Problem Statement

An EFB policy is a general explanation of how the EFB is expected to be used during flight operations and other activities. The purpose of a policy is to provide a framework within which procedures for using the EFB can be designed. Using a policy as the basis for procedure development will ensure that the resulting procedures are internally coherent and consistent with related procedures. Comparing a procedure to the underlying policy can aid in identifying discrepancies and conflicts with the policy. Pilots are more likely to conform to procedures developed from an explicit policy. A written description of the policy must be provided to all appropriate personnel.

Example(s)

An EFB policy could be similar to an operator's automation philosophy. It could describe the value the carrier expects to gain from the EFB and the role the EFB is expected to play in line operations (flight phases in which the EFB is to be used, etc.). It could also address expected changes in the duties of maintenance, dispatch, and other staff affected by the adoption of an EFB. To be complete, the policy must address each type of functionality that is supported by the EFB. An effective policy reflects the unique operational needs of the carrier.

Evaluation Questions
- Does the air carrier have an explicit policy that addresses the use of the EFB in line operations?
- How is the policy distributed to air carrier personnel?
- Are other policies affected by the introduction of the EFB?
- Does the policy adequately address each specific EFB application?

2.2.2 EFB Documentation

Training/Procedures Requirement(s)

> Existing Part 121 and Part 135 operator documentation must be modified, as necessary, to include information about the EFB.

Training/Procedures Recommendation(s)

> Adequate documentation should be provided to all EFB users describing the carrier's usage policy and providing guidance on how to use the EFB.

Problem Statement

The successful introduction of new equipment can be aided by the provision of adequate documentation that can be used for training purposes and as a resource for issues that may arise in the future. EFB information must be incorporated into existing documentation and could be provided additionally as a standalone manual.

Example(s)

Existing documentation may need to be modified to address EFB use within the larger context of flight operations. A separate EFB handbook may also be appropriate. In either case, this documentation could include the air carrier's policy on EFB use and an overview of the functionality supported by the EFB. The logic of the user interface could be described, together with the procedures for using the EFB under normal and non-normal conditions. Indications of a malfunctioning EFB and procedures for coping with a malfunctioning EFB are also important. Procedures that will be used to upgrade EFB software and content may be an appropriate topic. Finally, the document should list sources of additional information and help in using the EFB.

Evaluation Questions
- Is the documentation provided with the EFB sufficient?
- Did the air carrier incorporate EFB information into its current documentation?

2.2.3 User Feedback

Training/Procedures Good Practice(s)

> A formal process for gathering feedback from all personnel whose jobs are impacted by the EFB should be implemented for Part 121 and Part 135 operations.
>
> The introduction of the EFB into a fleet may benefit from small-group tryouts of the EFB, operating procedures, documentation, and training.

Problem Statement

Introducing a new piece of equipment into the flight deck requires changes to procedures, documentation, and training programs. Users are an important source of feedback for ensuring that effective changes are made. The feedback should be directed to a group whose purpose is to track issues and design features, so that design modifications can be requested.

Example(s)

A small-group tryout prior to full introduction into a fleet can be an effective way of evaluating changes to procedures, documents, and training—particularly on an early system prototype prior to finalizing the design. Rapid feedback can be obtained and the investment in staff training is minimized.

Once the EFB has been introduced, a formal process for accessing feedback from users can provide valuable information as users become experienced with the device. Pilots, check airmen, instructors, dispatchers and other personnel should be encouraged to submit their opinions and suggestions for improvements concerning procedures, techniques, documentation, problems occurring on the line, and training. Each submission should receive a formal response from an appropriate manager.

Advanced Qualification Program (AQP) carriers can use their data analysis methodology to gather evaluation information from all evaluation gates, including line checks and Line Oriented Evaluations (LOEs), to assess how well pilots are doing with the EFB.

Evaluation Questions

- Will the EFB be introduced using small-group tryouts? Does the tryout group include representatives from all user groups, including pilots of glass-cockpits and non-glass cockpits, maintenance personnel, and others noted above?
- Is there a formal process for gathering feedback about the EFB and its support?

2.2.4 EFB Training for Part 91 Operators

Training/Procedures Recommendations

> Part 91 operators/users should receive explanatory materials on how to use the device from the EFB manufacturer. These materials should be specifically designed for training, not just a system specification.

Training/Procedures Good Practice

> Hands-on training with a qualified instructor may be preferred for some aspects of using the EFB.

Problem Statement

Even "easy to use," well designed EFBs may be mysterious to new users at first, or they may have features that are mysterious even to experienced users. These difficulties may produce inefficient or incorrect use of the device, potentially affecting safety of flight.

Training may not be mandated for Part 91 operators, but a training guide should be provided. Hands-on training with a qualified instructor would be best.

Example(s)

Applications where interactive data entry is required, and where the resulting computations bear a direct effect on the safety of the flight may need more formal training than applications that do not affect the safety of flight directly. Also, training is more critical for EFBs that are used as primary sources of information.

Users who are transitioning to new aircraft may need only a differences-training program.

Evaluation Question(s)

- What materials and/or instruction are provided by the manufacturer on using the EFB?
- What materials and/or instruction are provided by the operator on using the EFB?

2.2.5 Initial EFB Training for Part 121 and Part 135 Operators

Training/Procedures Requirement(s)

> For Part 121 and Part 135 operators, training programs must be in place to ensure users have the necessary degree of competence in operating the EFB. Some home study may be an acceptable substitute for classroom study, provided all pilots demonstrate appropriate competence in operating the system prior to flight.
>
> EFB training must be designed based on an analysis of the minimum knowledge and skill requirements that must be met by users.

Problem Statement

The introduction of a new piece of equipment places additional demands on an already full training program. Safety considerations, however, require that a minimum acceptable level of proficiency be defined which all crew members must meet. The goal is to achieve this proficiency as efficiently as possible. Identifying the set of knowledge and skill requirements that constitute minimum proficiency, and using instructional techniques most appropriate for those requirements, can produce training that is both efficient and effective.

Training for airline crews may involve classroom or home study. Different types of training may be needed for completely new users (e.g., a new airline hire), users who are transitioning to new aircraft, and those who are undergoing annual continuing qualification training. The training programs will also vary based on the variety and criticality of applications on the EFB. Training for initial users may need to be integrated throughout their training on all aircraft systems, whereas training for more experienced users could be separate from training on other aircraft systems.

Training programs may have to be updated each time the EFB software and/or hardware is updated, unless the change is judged to be so small that an internal information bulletin (or equivalent) would be an acceptable substitute. The airline and its principal operations inspector (POI) need to agree on an acceptable training program.

Example(s)

The knowledge requirements will be determined by the functionality and design of the specific EFB. Some topics that might be covered include:

- Company EFB policy
- Procedures for operating the individual EFB applications, as well as procedures for using the EFB operating system, especially as related to using multiple applications at the same time
- Interactions (if any) with other aircraft systems, including all data transactions
- Known quirks of the EFB
- Procedures for updating content and upgrading the software
- MEL status, including paper or other backup
- Troubleshooting procedures

Skill requirements will involve proficiency in performing all functions for which the EFB is to be used. If initial EFB training is accomplished as a part of qualification training, EFB use can be trained within the context of procedures for operating the aircraft.

Evaluation Question(s)

- Does the air carrier's initial EFB training address all knowledge and skill requirements?

2.2.6 Evaluation Process for Part 121 and Part 135 Operators

Procedure/Training Requirement(s)

> Evaluation of EFB proficiency during initial training for Part 121 and Part 135 operators must address both the knowledge and skill requirements.
>
> Recurrent training for this group must include appropriate use of the EFB within the context of normal and non-normal procedures.

Problem Statement

Appropriate evaluation of EFB proficiency is key to ensuring that pilots achieve minimum proficiency during initial EFB training and maintain that proficiency during line operations as evaluated through both line checks and recurrent or continuing qualification training. EFB evaluation should be consistent with the carrier's EFB policy and standard operating procedures.

AQP training, as well as some Part 121 programs, uses a series of evaluation gates, which supports regular testing and, therefore, early detection of weaknesses in a pilot's knowledge or skills. The number and types of evaluation gates will depend on whether the EFB training is conducted apart from qualification or continuing qualification training or is integrated into existing programs. In either case, evaluating proficiency on the EFB as a part of initial EFB training involves both a knowledge and a skill component.

Example(s)

Evaluating EFB proficiency as a part of recurrent EFB training can take place within the context of appropriate evaluation gates. First-look and maneuver validation are appropriate for evaluating the use of the EFB during the performance of procedures. The use of the EFB as a decision tool and workload management aid is better addressed during line oriented evaluation (LOE).

Evaluation Question(s)

- Does the carrier's initial EFB training support evaluation of both knowledge and skill requirements?
- Does the carrier's recurrent or continuing qualification training include evaluations of EFB proficiency during all appropriate evaluation gates?

2.2.7 Fidelity of EFB Training Device

Procedure/Training Requirement(s)

> The level of fidelity of the EFB training device must match training requirements.

Problem Statements

The issue of what constitutes sufficient simulator fidelity for training and evaluation purposes has been of interest to the air carrier community for many years. Historically, full fidelity was assumed to be the ideal. However, it is now known that the degree of fidelity that is required depends upon the specific training goals. Training device fidelity is also an issue for EFBs and must also be driven by training requirements.

The cost of a fully ruggedized EFB that meets all certification requirements for use on the flight deck may be prohibitive when the intended use of the device is for training purposes. Fidelity requirements must be driven by the training requirements during each stage of the training process. A clear definition of the training goals must drive decisions concerning the required level of fidelity.

Example(s)

For example, users may be able to achieve many knowledge requirements by using a standard laptop computer that runs the EFB software. The software itself must be identical with that used on the flight deck but the hardware may not need to be.

Training the user on the procedures for operating the device will require that the hardware interface be identical to the device used in flight, but the training device need not be ruggedized.

Evaluation Questions

- Does the EFB training device that is used during each phase of training provide the required degree of fidelity?

2.2.8 Ensuring Data Integrity

Training/Procedures Requirement(s)

> Procedures must be in place to ensure that the data upon which flight information is based comes from an appropriate source.
>
> Databases that are loaded onto an EFB must be checked by appropriate methods to ensure that they are accurate, up-to-date and uncorrupted.

Problem Statement

EFBs may provide information that is critical for flight safety, so that information must be based on correct, uncorrupted data, or else the integrity of the information is questionable.

The flight crew is ultimately responsible for ensuring that any user-entered data comes from an appropriate source. Procedures must ensure that the crew does not forget to review or re-enter any user-entered data.

If the EFB is a "read-only" device, it may only be necessary to check the databases when they are installed or updated.

Example(s)

While parked at the gate, a maintenance person may wish to browse a manual that he/she knows is on the EFB. Having forgotten their own copy, they may use the EFB to look up the information while the crew is out. When the crew returns they may do one of the following sample procedures for reviewing the EFB data:

(1) clear all user-entered data and re-enter it themselves

(2) review and check all user-entered data (which has been highlighted by the EFB software)

(3) review all highlighted data, which was perhaps entered by a remote source such as Flight Dispatch, and re-enter all locally entered data

If a calculation is based on the user-entered data, another approach would be for the pilot to pass through all the calculation steps, even if no data entry is required; i.e., to review all the raw data before the calculations are performed.

Evaluation Question(s)

- What are the procedures for ensuring that EFB data comes from an appropriate source?
- What are the procedures for ensuring that the EFB databases are accurate, up-to-date, and uncorrupted?

2.2.9 Use of Hand-Held EFBs

Training/Procedure Recommendation

> While in use, EFBs that are not held in a structural cradle should not be routinely placed such that they obstruct access to other critical flight controls/displays.

Problem Statement

EFBs that have no designated location while they are in use, such as a cradle, may obstruct access to other displays/controls. This problem will be especially pronounced if the EFB is physically large enough, relative to the size of the flight deck, so as to be difficult to moved about quickly and easily.

Example(s)

An EFB might be placed on the pilot's knee during takeoff (perhaps using a securing strap). If so, the pilot must insure that he/she has full control authority, i.e., that the yoke can be pulled back completely, without the EFB getting in the way.

Evaluation Question(s)

- Does the pilot have adequate access to critical flight controls/displays when the hand-held EFB is in use?
- What are the procedures and/or training for routine placement of hand-held EFBs?

2.3 Equipment

2.3.1 Input Mechanisms

Equipment Recommendation(s)

> The EFB display and control hardware should meet the requirements given in the January 1996 FAA Human Factors Design Guide for Acquisition of Commercial-off-the-shelf Subsystems, Non-Developmental Items, and Developmental Systems—Final Report and Guide (Section 7).

Equipment Issue(s)

> In choosing and customizing input mechanisms, such as keyboards or cursor-pointing devices for an EFB, designers should consider the type of entry to be made and flight deck environmental factors that can affect the usability of that input mechanism, such as turbulence. Each of these input mechanisms has parameters that should be optimized for use in a flight deck environment.

Problem Statement

The user will interact with EFB functions via some input mechanism, such as a track ball, touch pad, rotary knob, keyboard, or soft keys. The input mechanism should be matched to the complexity of the entries to be made, and the entries to be made will vary by the function that is being performed. If the input mechanism is poorly matched to the task, not only are errors more likely, but the task with will take considerably more time to complete, and users will become frustrated, and potentially even distracted from higher priority tasks.

Example(s)

The simplest types of input may be to select a text hyperlink on an electronic document, using a cursor control. More complex input, such as numeric data for performance calculations may require the use of a simple, or full-featured keyboard. If the EFB does not support entry of free text, a full keyboard may not be necessary.

As an example of optimizing input mechanisms, consider the case of a touch screen input mechanism; the size of the active areas may need to be larger in a flight deck environment than it would be in a stable environment to promote accurate data entry. This will in turn impact the size of the display, and the size of the unit itself.

Evaluation Question(s)

- Does the EFB input mechanism meet the requirements given in the January 1996 FAA *Human Factors Design Guide Section 7?*
- Can the user easily perform the most common types of data entry?
- Can crews use the input mechanism accurately and reliably for the least common types of data entry without an unusual level of skill, patience, or practice?
- Can the user position the pointer/cursor mechanism (if any) reliably and repeatedly under all flight conditions (e.g., turbulence, darkness)?

2.3.2 System Error Messages

Equipment Requirement

> If an application is fully or partially disabled, this status must be clearly indicated to the user with a positive indicator.

Equipment Recommendation(s)

> The EFB system error messages should meet the requirements given in the January 1996 FAA Human Factors Design Guide for Acquisition of Commercial-off-the-shelf Subsystems, Non-Developmental Items, and Developmental Systems—Final Report and Guide (Section 8).

Problem Statement

There are many reasons why systems fail to operate as expected. For example, functions that require external data may fail when the that data is not received, producing a partial failure. A total failure may occur if there is a hardware fault.

The user should be aware of the system status at all times. Without a clear indication of the EFB status, the user may make decisions based on outdated, incorrect, or incomplete information.

A positive indicator of failure (e.g., a warning light or message that appears upon a failure condition) is clearer and more noticeable than a negative indicator (e.g., a warning light or message that turns off.)

Example(s)

If the EFB application is integrated with other flight deck systems and this connection fails, the user must be aware of this failure. For example, an EFB electronic checklist may be designed to bring up a non-normal condition checklist upon encountering such a condition. However, if the EFB has lost its connection, it may not know of the non-normal condition and the pilot may miss an associated checklist because he/she presumed that the checklist function was working correctly.

As another example, if the EFB provides real-time information, such as weather or air traffic clearances via some type of data link system, the user should be notified if there is a problem with the data link that precludes normal display of the data. If the link is down completely, and there is no data to display, this must be distinguished from the case where there is a blank screen because, for example, there is no traffic or precipitation in the selected region. If the link is operational in a degraded mode (e.g., the data rate is half of the normal rate so that the data is refreshed less often) this must also be brought to the crew's attention. For data link services, this requirement to notify the crew of system errors should be consistent with the Minimum Operational Performance Standards for that service and with AC-140 (see reference list).

Evaluation Question(s)

- Are partial or full failures of the EFB clearly annunciated with positive indicators?
- Do the EFB system error messages meet the requirements given in the January 1996 FAA *Human Factors Design Guide Section 8?*

2.3.3 Compatibility Across Applications on the EFB and Use of Style Guides

Equipment Recommendation(s)

> All applications on the EFB should be designed in conformance with the style guide for that system if one is available. Note: Style guides are available for industry standard operating systems, such as Windows™ (see reference list).
>
> If an industry style guide is not available, developers of EFB hardware and operating systems should write a user interface style guide for application developers to explain the user interface principles and conventions for that system. These style guides should include standard practices on performing common actions (e.g., opening and closing documents, selecting and editing text, or printing) and standard user interface elements (e.g., windows, menus, dialog boxes, and system alerts). Conventions for the behaviors of the input mechanism (e.g., "single" or double-clicking), use of color, and icons/graphic elements, standard shortcut conventions, and navigation methods should also be documented.
>
> The user interface should be designed in accordance with appropriate industry guidance materials (e.g., from the Society of Automotive Engineers (SAE) and RTCA).
>
> The user interface should be designed with a consistent set of controls (e.g., buttons) and graphic elements (e.g., icons, windows, and menus). Controls that are used for different purposes should be visually distinct from one another. Functional properties of graphic elements and controls should follow standard personal computer conventions.
>
> There should be a consistent convention for use of color and other formatting (e.g., use of underlining) across the EFB applications. There should be a standard help facility convention across the EFB applications.
>
> Specific common actions that are allowed on multiple applications(e.g., launching or exiting an application, or selecting a hyperlink) should be performed in the same manner.

Problem Statement

To date, EFB prototypes have all used software customized for them by a single vendor. But in the near future, customers may purchase EFB software from different vendors, much as today one can buy software written by different companies that runs on a single desktop personal computer. If the applications are not written using the same types of graphical user interface elements and conventions, users will take longer to learn the individual software applications. In addition, users will be more likely to make errors and become frustrated with all of the applications, not just the ones that are different from the standard applications provided by the original system developers.

Manufacturers who are constructing their own EFB operating system and library of graphical routines should keep in mind that internal consistency of the system is critical to user acceptance and system usability. Style guides can greatly improve the internal consistency of applications.

Example(s)

Software style guides are common in the personal computer industry (see reference list). Manufacturers of operating systems specify the conventions they use in these style guides so that application developers can build consistent graphical user interfaces. Another advantage of style guides is that they can shorten the time to develop applications by giving software developers guidance on the correct user interface design methods.

Evaluation Question(s)

- Does the EFB software conform to any existing style guide for that system?
- Did the manufacturer of EFB hardware write their own style guide?

2.3.4 Multi-Tasking

Equipment Recommendation(s)

> If the EFB is able to run more than one application at a time, the user should be able to find out which flight-deck applications are running and switch to any one of these applications easily. When the user returns to an application that was running in the background, it should appear in the same state as when the user left that application.
>
> The responsiveness of any individual application (e.g., time to process user input or complete computations) should not suffer when all other supported applications are running.
>
> To exit applications that have pending activities, the user should deal with these activities, either completing them, or overriding them with an extra confirmation step. The user should be allowed to switch applications even when there are pending tasks in the original application. The currently active application should be clearly indicated.
>
> Some EFBs will support applications that are not directly related to flight tasks (e.g., word processor, or game). In order to discourage inappropriate use of non-flight-related functions, the system should warn the user that these applications are not recommended for use in flight, and ask for confirmation to proceed when they are launched. Where appropriate, airline/operator policy should specify which applications are approved for use in flight.

Equipment Issue(s)

> For each application running on an EFB, the designer should consider whether summary information should be available in the foreground, without requiring the user to activate that application. The operating system should support the ability to view basic information about an application without activating that application.
>
> Also, the designer should consider whether an application should run automatically (i.e., the user may not have to launch the application manually) to insure that it is always active.

Problem Statement

The user may need to switch between multiple active applications quickly. If switching is cumbersome, time-consuming, or error prone, the pilot's workload will increase In order to avoid switching between applications often, it may be useful to be able to view some basic information without having to activate the background application.

Example(s)

One way to manage multiple applications is from a soft-key "main" menu that has a list of all available applications. Users should be able to get to this main menu with a single key press from any point in the software. The main menu should appear on the default starting screen.

Some applications, such as an ECL, or traffic display, could start automatically when the EFB is turned on since they may be useful immediately. These applications should also close automatically when the EFB is turned off. The user should still be able to close them manually, with a warning to finish pending tasks and user confirmation. For example, if the user attempts to close the ECL application, and it is keeping track of one or more incomplete checklists, the user should deal with checklists before exiting the ECL.

For an ECL that is running in the background, it may be useful for the user to know the name of the currently active checklist, if any, in the foreground. This technique is especially helpful when the background application has time-dependent information (e.g., traffic information). For traffic information, for example, a running count of the number of other aircraft within a user-specified distance, such as 10 miles, may be useful to have in the foreground.

Evaluation Questions

- Can the user switch between currently active applications with a single step? Does the background application reappear in the same state that it was left at?
- Is there an extra confirmation step required to get to any applications that are not flight related?

2.3.5 Responsiveness of Application

Equipment Requirement(s)

> The system must provide feedback to the user when a user input is processed.
>
> Alphanumeric inputs must be drawn on the display within 0.2 seconds. If the system is busy with internal tasks that preclude immediate processing of user input for longer than 0.5 seconds (e.g., calculation, self test, or data refresh), a "system busy" indicator (e.g., clock icon) must be displayed to show the user that the system is occupied and will not process inputs immediately. (Source: 1996 SAE Aerospace Recommended Practices for Data Link Systems (ARP 4791).

Equipment Recommendation(s)

> Feedback for user input should vary based on the type of input.
>
> If an internal task takes more than a few seconds to complete and user entries are not processed immediately while the task is in progress, a progress indicator should be displayed to show the user how much of the internal task is complete so that he/she has a sense of when entries will be processed.
>
> User entries that are made while the system is occupied should be stored and processed as soon as resources are available.

Equipment Good Practice(s)

> Many industry applications draw an unobtrusive busy indicator *whenever* the application is occupied. Then, if the system is occupied only briefly, the busy indicator is hardly noticeable, but if the application is occupied for a longer time, the user will notice it as necessary.

Problem Statement

Immediate user feedback must be provided so that the user is aware that the system has received and accepted his/her input right away. Without feedback, the user may try to re-enter inputs multiple times in a short interval, and he/she will become confused when the system eventually acts upon the multiple inputs.

If user entries are not being accepted, this status must be displayed so that the user does not attempt to make entries that are then discarded by the system. Busy indicators are necessary to show that the system is occupied with internal tasks. A progress indicator is useful so that the user can judge how much longer he/she has to wait before entries will be accepted.

Example(s)

Some examples of feedback for user inputs include the following:

- If the user enters a character, the character is drawn on the screen as feedback.
- If the user makes an unacceptable input, such as entering a character when none is applicable, then an audible error beep could indicate that the entry was received, but inappropriate.
- If the user manipulates the cursor pointing device, its location should be updated and redrawn.
- If the user select an action from a button, the feedback could be the start of that selected action.

Busy indicators are often shown by changing the shape of the pointer on industry standard GUIs. Typical indicators look like clocks. Progress indicators can take many forms. Some common forms include graphical progress bars that are filled in as the task is completed. Text progress indicators are also acceptable. For example, a text message could state how many tasks have been completed, and how many remain to be completed.

Evaluation Question(s)

- Is feedback provided for all types of user inputs within an appropriate time?
- Are busy indicators displayed when processing of user input is delayed?

2.3.6 Soft Keys

Equipment Recommendation(s)

> Soft keys should provide tactile feedback to the user when pushed.
>
> Note: Soft keys are physical buttons whose actions can be reassigned via software.
>
> The software should be designed to filter out multiple entries from a soft key if they occur too closely together for the user to have intentionally made them as separate actions.
>
> The labels used to identify the action associated with a soft key should be clear to the user and brief. The labels should also be used consistently throughout the software.
>
> Soft keys may be used to select one of several available actions. When the same set of options is accessed from different points in the software, designers should make sure that the same function appears on the same key at all times.

Problem Statement

Soft keys are useful because they can be reassigned to a variety of actions within software. However, they can also be difficult for the user to use accurately if designed poorly. Poor soft-key design can increase pilot workload by requiring the pilot to check his entries more carefully, and by having to correct incorrect entries more often. Tactile feedback can help ensure that the pilot is aware of his/her entries.

When the soft keys are consistently mapped with a particular action, users can associate that key's location with that action, making the software easier to use.

Example(s)

Also, multiple entries are often registered by the hardware when a user holds down a button longer than expected. A good rule of thumb is to discard multiple entries that occur within 300 milliseconds of each other.

Evaluation Questions
- Do soft keys provide tactile feedback?
- Are soft keys labeled consistently?
- Are inadvertent multiple entries discarded?

2.3.7 Anchor Locations

Equipment Recommendation(s)

> It should be easy for the user to move from any location in the EFB to an anchor location, such as the main menu, or home page for that application. The user should be able to move from the anchor location to other EFB functions easily.
>
> There should be an anchor location from which the user can move between EFB applications if more than one application is supported.

Problem Statement

In order to be useful to the pilot, the information on an EFB needs to be accessible. In order to be accessible, the pilot must know how to get to any desired data quickly, starting from any other point in the EFB. One of the standard ways to help the user orient him- or herself is to provide an anchor location such as a top-level main "menu", or a "home page." When these anchor locations are easy to access, the user can jump from one place in the EFB to another quickly, easily, while staying oriented.

If there are no anchor locations, the user is more likely to get disoriented and have trouble moving from one place to another in the EFB.

Example(s)

In a full graphical user interface, the "desktop" could be the main anchor location. It would be easy to return to the desktop from any where in the EFB functionality. From the desktop, the user could determine what EFB functions were running, and switch to any other function.

In a soft-key menu-based user interface, a top-level menu could be the main anchor location. The user would be able to go to that top-level menu by pushing a single, dedicated button at any time. Individual EFB applications, such as an electronic document viewer, or a traffic display could have their own anchor pages, from which the user could access any of the functionality of that particular application, such as opening a document, or viewing the graphical traffic display.

Evaluation Questions

- Are there anchor locations in the EFB software?

2.3.8 Legibility of Text—Character Issues

Equipment Requirement(s)

> The EFB must use a highly legible typeface that enables the user to quickly and accurately identify each character. In particular:
> - Individual characters should not be easily confused with other characters.
> - Characters with stroke widths should provide sufficient contrast between the character and the background.
> - Characters should be drawn with constant stroke widths.
> - Use of slanting or italic characters and upper case text should be avoided.

Problem Statement

In order for information to be quickly and accurately understood by the user, each typeface that is used by the EFB must be highly legible. Legibility is affected by such factors as the shape of each character, the width of the strokes that form each character, and the avoidance of italic and upper case text.

Example(s)

Sans serif typefaces do not use small horizontal strokes at the top or bottom of characters (e.g., "h" or "y"). Serif typefaces do have small horizontal strokes at the top or bottom of characters (e.g., "h" or "y"). Sans serif typefaces are typically more legible than serif typefaces on a computer screen. If a serif typeface is used, a higher screen resolution may be necessary to achieve comparable legibility.

The similarity of individual characters also affects legibility. In order to achieve a "family" appearance, some typefaces use characters that appear quite similar. Characters which are most likely to be confused are "P" and "R"; "B," "D," and "E"; "G," "O," and "C;" "l" (the letter) and "1" (the number); and "Z" and "2."

Upper case text is more difficult to read and, therefore, should be used sparingly. It should not be used for emphasis. Slanted or italic text should be avoided for the same reason.

Evaluation Questions
- Are individual characters easily recognized for each typeface that is used?
- Does the typeface use strokes with sufficient and constant width to enable each character to stand out against the screen background?
- Is upper case and italic case avoided?

2.3.9 Legibility of Text—Typeface Size and Width

Equipment Requirement(s)

> The EFB must use a typeface size that is appropriate for the viewing conditions and the criticality of the text.

Equipment Recommendation(s)

> The January 1996 FAA Human Factors Design Guide for Acquisition of Commercial-off-the-shelf Subsystems, Non-Developmental Items, and Developmental Systems—Final Report and Guide recommends the following:
>
> —Using a typeface height of at least 1/200 of the viewing distance.
>
> —Using a larger typeface size for text that can be expected to be read under low-visibility conditions (e.g., some emergency checklists).
>
> —The ratio of character height to width should be:
>
> —At least 1:07 to 1:09 for equally spaced characters and when lines of 80 or fewer characters are used.
>
> —At least 1:0.5 if more than 80 characters per line are used.
>
> —As much as 1:1 for inherently wide characters such as "M" and "W" when proportionally spaced characters are used.
>
> —As much as 1:1 for inherently wide characters such as "M" and "W" when proportionally spaced characters are used.
>
> If these guidelines are not met, there should be a sound basis for deviation.

Problem Statement

Typeface size is a critical determinant of the ease with which text can be read. The variety of lighting conditions under which the text must be read must be considered. Equally important is the type of information being conveyed. Critical information that must be read under potentially low-visibility conditions must be displayed using a larger height in order to ensure that the users can quickly and accurately read the information. Equally important is typeface width. Narrow characters can be more difficult to read.

Example(s)

The minimum typeface size that is used must support legibility under a wide range of lighting conditions. EFB information must, therefore, be presented using a minimum typeface height of that is 1/200 of the viewing distance. For a viewing distance of approximately 31 inches, the characters must be at least .16 inch high. A larger size may be required for some applications. In particular, emergency checklists that will be used under low-visibility conditions, such as checklists used for smoke-related conditions, must use a larger size.

The recommended width of typeface characters is defined as a ratio of character height to width. Wider characters (1:0.7 to 1:0.9) are needed for equally spaced characters and lines of 80 or fewer characters. Narrow widths (a minimum of 1:0.5) may be used for lines of text having 80 or more characters. Naturally wide characters ("M" and "W") will require a ratio as high as 1:1.

Evaluation Questions

- Is the typeface size legible under normal viewing conditions?
- Is the typeface size used for emergency checklists and other critical text that can be expected to be used under low-visibility conditions adequate?

2.3.10 Legibility of Text—Spacing for Readability

Equipment Requirement(s)

> In order to make text easily readable:
> - Use a horizontal spacing between characters of at least 10 percent of character height.
> - Use spacing between words of at least one character when using equally spaced characters or the width of the capital letter "N" for proportionally spaced characters.
> - Use a vertical spacing between lines of at least two stroke widths or 15 percent of character height, whichever is larger. Vertical spacing begins at the bottom of character descenders and ends at the top of accent marks on upper case characters.
> - Use line lengths that are appropriate for the type of text (e.g., checklists)

Problem Statement

Text displayed by the EFB must be readable as well as legible. Readability is primarily concerned with enabling the reader to easily recognize words and keep the reader's eye from unintentionally skipping to another line of text. Readability is determined by the spacing between individual characters and words, between lines of text, and by the length of the line of text.

Example(s)

Appropriate vertical and horizontal spacing must be used between characters to support the user's eye in recognizing the boundaries between characters, and between words. Tight spacing between characters can cause the characters to run together while loose character spacing makes the boundaries between words less detectable. Spacing between characters must be at least 10 percent of character height while word spacing must be at least one character spacing for equally spaced characters or the width of the capital letter "N" for proportionally spaced characters.

Spacing between lines of text must be the larger of two measures: at least two stroke widths or 15 percent of character height. Vertical spacing is measured from the bottom of character descenders (that part which descends below the text line as seen in the lower-case letter "y") to the top of the accent marks (if used) on upper case characters.

Long lines of text can cause the eye to jump to the next line. Line length is especially critical for checklists where a large gap between the challenge and response items may cause the reader to pair a response item from a different line.

Evaluation Question(s)

- Does the EFB use a horizontal spacing between characters and between words that clearly indicates boundaries between words but does not cause individual characters to blur with each other?
- Is there sufficient vertical spacing between lines to help the user's eye avoid skipping to the next line?
- Are the line lengths used appropriate for each type of text?

3 Electronic Documentation

The first function proposed for EFBs was support of electronic versions of the document in a typical airline pilot's flight bag, such as reference manuals. The goal is to convert paper documents into electronic documents while retaining, if not enhancing, overall readability and access to the data. For example, electronic documents could be cross-linked to make it easy for users to read about a single topic that is addressed in more than one document. Electronic documents are also expected to be easier to update and distribute.

Both paper-based and electronic document systems require information to be logically organized and structured to ensure the user can access and use the information with minimal effort. Poorly organized and structured paper-based document systems will not be enhanced by simple conversion to an electronic media; such information must be reorganized/structured before conversion.

Certain paper-based attributes such as tables of content, indices, and cross references can be significantly enhanced by automatic linking in an electronic media (i.e., hypertext links). In addition, electronic media can allow full text searches which enable the user to access information across multiple documents.

In Section 3.1, the type of documents that are being considered here are clarified and the many options that electronic documents might or might not support are reviewed. Considerations for electronic documents begin in Section 3.2, with General Issues. Layout and Appearance issues are presented in Section 3.3. Navigation and Search issues are presented in Section 3.4. Finally, optional (generally advanced) electronic document features are presented in Section 3.5.

3.1 Background

3.1.1 Type of Documents Addressed

Pilots carry at least three types of documents in their flight bag: manuals, checklists, and navigation publications. Only manuals are covered in this section. Checklists are covered in Section 4 and navigation publications will be covered in a later draft (Version 2) of this document. The manuals considered here include:

- Pilot's Operating Handbook (POH), which contains information on the aircraft and its systems
- Flight Operations Manual (FOM), which contains airline policies and procedures, including emergency procedures
- Airport Analysis and Aircraft Performance Manual, which contains information about specific airports and runways
- Minimum Equipment List (MEL) and configuration deviation lists, which contain information about operational restrictions stemming from limitations in case of partial or full failure of various aircraft instrumentation and systems

The FAA mandates that these documents be on board the aircraft, approves or accepts their content, and reviews any changes. In the future, other similar documents, such as the Aeronautical Information Manual, the Federal Aviation Regulations, and Aircraft Maintenance Manuals, could be available on an EFB.

These flight manuals are primarily reference documents that are used relatively infrequently to find specific information, although they are sometimes used for studying topics in depth. The manuals are updated relatively infrequently (i.e., over a period of days, not hours). They are not used interactively in the way that electronic charts, or electronic checklists will be used, where the pilot routinely customizes the display, or enters data. Also, these manuals are not usually accessed in time-critical situations.

3.1.2 Features of Electronic Documents

Electronic documents may vary greatly from one another. They may vary in how the document is displayed, whether the view is customizable, how the user enters information and commands, how the user enters text (if at all), and what level of support there is for multimedia. All of these features are affected by the building blocks used to create electronic documents (i.e., the library of software routines). These building blocks are usually built into the operating system of the EFB device, though they are not required to be.

Advanced and basic operating systems are compared in Table 3-1 below. In general, advanced operating systems support advanced document features, but this is not always the case. For example, if the operating system supports windows, the user usually is given the option to reposition and resize the windows. However, it is best to examine each implementation of electronic documents feature by feature.

Advanced and basic electronic document features are compared in Table 3-2 on the next page. Display and input hardware limitations that may affect the user interface of electronic documents are also listed in Table 3-2.

Another independent factor that affects electronic document functionality was already mentioned in Section 1: is the EFB integrated with other flight deck systems? If it is, then electronic documents that are related to a given system condition (e.g., failure of an item) could be suggested to the crew from built-in logic. In this case, the electronic document functionality is a decision support tool, not just a data-access tool.

Two examples illustrate the spectrum of electronic documents. First, consider a simple display system that shows unformatted text and static graphics displayed in a single window or frame. This system might marginally serve as in information repository, but would probably not be significantly better than its originating paper document, and could be worse. Without information being logically and visually "chunked", neither paper nor electronic system would probably satisfy pilots needs.

A second example would be converting an existing, well-crafted paper document to an electronic page-based display system (e.g., Adobe Acrobat). This would preserve the original document's organization and structure and additionally provide benefits of electronic media (e.g., hypertext linking, full-text search). Some drawbacks include certain topographical considerations such as typeface optimization for screen presentation, and screen size display capability.

Operating System	<u>Advanced</u>: Industry standard (e.g., Windows, Macintosh, or Unix) or equivalent proprietary system. The operating system supports a graphical user interface with windows and menus. There are standardized dialog boxes with standard interface elements, such as text entry boxes and command buttons. There are standardized dialog boxes for displaying system status and alerts, which are compatible with other types of dialogs boxes. The operating system is typically designed for use on personal/business computers outside of the aviation environment. It can run more than one software application at a time. Text formatting (e.g., changing font color, size, or underlining) is fully supported.
	<u>Basic</u>: Compatible with industry standard operating system to some degree (e.g. Palm OS or DOS). The basic operating system could be more or less compatible with industry standard. For example, the graphical user interface could be limited, or unsupported. Basic dialog boxes may be supported, such as modal system alerts or simple user entry dialogs (e.g., with yes/no responses), but complex user entry dialog boxes are not supported. Typically, these systems have soft keys, which are hardware buttons whose functions are configurable. Text formatting could be limited or unavailable.

Table 3-1. Differences between advanced and basic EFB operating systems.

Display Area	<u>Advanced</u>: Documents are displayed in windows that can be repositioned on the screen and resized. The windows support scrolling text and color. Page layout can be formatted similar to a paper page on the window. More than one window/document can be open at a time and overlapped, with one window covering part of another window. May allow the user to customize the display in some ways. For example, windows could be resized or repositioned, or font properties could be modified if allowed.	
	<u>Basic</u>: Either (a) or (b). Neither typically allows the user to customize the display. (a) Regions/Frames. The screen area has selectable regions/frames. The regions may support scrolling, but cannot be repositioned or resized individually. The basic display area does not support overlapped display regions. (b) No Regions/Frames. The screen area does not have selectable regions/frames. Scrolling, if supported, affects entire screen contents.	
General User Input	<u>Advanced</u>: Users can point to any area of the screen and activate that point (e.g., by using a mouse, touch pad, pen input). If that area is active, a change will occur; for example, a menu may drop down or pop-up, or a cursor may be moved to that position. Users may select actions via graphical (pop-up or pull-down) menus. The list of menu options is out of view until called up by user. The on-screen cursor may allow the user to select regions of text, for example, for printing or copying. The cursor could also be used to activate hyperlinks or to activate pop-up areas for definitions, help, or related information on a topic. There may be standardized dialog boxes with standard interface elements, such as text entry boxes and command buttons. There are standard practices for designing dialogs for confirmation of important actions and collecting user-entered data. Dialog boxes could be modal (i.e., no other user actions are accepted until dismissed) or modeless (i.e., other user actions are processed normally while the box is visible).	
	<u>Basic</u>: Soft keys are used to select menu items, activate commands, and navigate between menus and screens. The same buttons could move the cursor or scroll text. The user must always select from a visible list of alternative actions. (There are no pop-up or pull-down menus.)	
Text Entry	<u>Advanced</u>: Users can enter free text easily (e.g., through a keyboard, or even hand-writing recognition software). The free text capability could be used, for example, to enter annotations, or to enter key words for search.	
	<u>Basic</u>: Either (a) or (b) (a) Keypad-style text entry. Text entry is possible but usually cumbersome because more than one key press could be needed to enter a specific letter. For example, letters could be entered with 10 push buttons, configured as on a telephone. (To enter a "B", press the "1" key, which corresponds to "A", "B", and "C", twice.) No text entry supported.	
Multimedia Support	<u>Advanced</u>: Includes support for audio and visual animation, such as video clips, and/or sound recordings.	
	<u>Basic</u>: Limited or no support for audio and visual animation. For example, the audio may be limited to a system beep, or a few system sounds.	

Table 3-2. Comparison of advanced and basic electronic document features.

3.2 General

3.2.1 Consistency of Logical Structure Between Paper and Electronic Documents

Equipment Requirement

> The logical structure of an electronic document must be consistent with the hard copy version of that material, if a hard copy exists.

Equipment Issue

> When converting paper manuals into electronic manuals, it may be possible to streamline the electronic document by deleting content that is not relevant to the aircraft in which the EFB will be used. When content is considered for deletion, designers should ensure that that data are not necessary in any unusual circumstances. Also, if entire sections of text are deleted, they should still be called out by heading, so that consistency with the more complete paper document is maintained.

Problem Statement

Pilots are very familiar with the logical structure of hard copy versions of required documents. They use this background knowledge when searching for information through a table of contents or index. In cases where the operator determines that electronic documents are better designed for pilot use than the paper versions, it is appropriate to update the paper manuals. Consistency between the two documents' logical structures is required, but evolution towards a better structure for all versions is encouraged.

Example(s)

The section headings and section numbers for any reference manual should be the same in both the electronic and paper versions of the document. It may not be necessary to keep page numbers on the electronic version, as long as the content can be referenced in other ways, such as through section headings.

An example where customization of electronic documents could be useful is when there are multiple aircraft models or series that are very similar. For example, there is one paper manual for both Boeing 757 and 767 aircraft. Within this document, there are notes next to any material that applies to only one of the two types. An electronic version of this document could be configured to display only the material relevant to the aircraft model of interest.

Evaluation Questions

- Is the logical structure of the electronic document consistent with any hard copy version of that material?

3.2.2 Consistency of Electronic Document User Interface

Equipment Recommendation(s)

> The electronic document user interface should be designed consistently with the operating system style guide.
>
> The electronic document user interface should also be internally consistent.
>
> Text color and formatting should be used consistently.
>
> Consistent heading styles and formats should be used to highlight warnings and cautions.

Problem Statement

As noted in Section 2.3.3 (Compatibility Across Applications on the EFB and Use of Style Guides), internal system consistency is critical to user acceptance and system usability. Consistency should be maintained both within and across applications.

The electronic document user interface should be consistent with regard to use of windows and/or frames(if implemented), use of shortcut keys, labels and terminology, and use of graphical elements such as command buttons and scroll bars. That is, the controls and visual display of an electronic document should be similar, regardless of the which document is displayed.

Example(s)

For example, unvisited hyperlinks could be identified as underlined text of one color, and visited hyperlinks could be underlined text of a different color.

Change bars, or other text formatting could be used to identify recent changes in the text.

Heading styles and formats could be used to highlight specific types of text, such as "special notes."

Evaluation Questions
- Is the electronic document user interface designed consistently with the operating system style guide?
- Is the electronic document user interface internally consistent?
- Is text color and formatting used consistently?

3.2.3 Training Needs

Training/Procedures Recommendation(s)

> Users should receive training on how the logical structure of paper documents matches that of the electronic documents.
>
> Basic procedures for moving through an electronic document should be addressed in training. These includes how to choose a document for display, displaying a selected section, and moving between pages in the same document segment.

Training/Procedures Issue

> Users are likely to need more training to work with electronic documents that support advanced features.

Problem Statement

To support compatibility between the paper and electronic versions of the same documents, electronic documents must implement the logical structure of the paper document (see 3.2.1). Users should understand that the electronic documents follow the same logical structure as the paper documents.

Users should also understand how to access and display different segments of the electronic document(s) available on the system.

More complex user interfaces allow more flexibility and support more features, but they can also be confusing and frustrating for users. In particular, it is easier for users to end up in an undesired configuration without knowing how to get out of that configuration. For this reason, it is expected that more advanced electronic document applications will require more training. Not only is the training time likely to increase, but the training program will also have to be designed more carefully.

Example(s)

Electronic documents may have a table of contents similar to the paper document. The user could recognize the document's logical structure from this table.

More sophisticated functionality will require sufficient training to enable users to benefit from the functionality and protect them from getting distracted or frustrated. For example, users will need to understand how to customize documents if customization is allowed. More complex navigation and search techniques will also add to the training curriculum.

Evaluation Questions

- Does the training program provide adequate instruction in how to navigate through an electronic document?
- Does the training program adequately cover how to use the advanced features of the electronic document? Do users know how to avoid using advanced features if they so choose?

Version 1 (9/28/00)

3.2.4 Speed of Loading Data

Equipment Recommendation(s)

> Users should not have to wait more than a few seconds for the display to refresh with new content.

Problem Statement

While access to flight manuals is not always a time-critical task, it is important that the documents are loaded quickly when requested. If the files are stored locally, this is not likely to be a problem. However, if files are accessed remotely, for example, through a server, or perhaps down loaded via data link, delays may become an issue. Also, some types of information, such as figures and tables, may be slower than average to load.

Example(s)

If the flight crew opens a new manual electronically, the first page of the new text should be visible within a few seconds. As the user moves about the document to different sections with different types of content, each screen refresh should take no more than a few seconds.

Evaluation Questions

- Is new content, including complex data, such as figures, or tables, displayed within a few seconds?

3.3

3.3 Layout/Appearance

3.3.1 Visual Structure

Equipment Issue

> The visual structure of a document can significantly affect its readability and ease of comprehension. Factors such as font choice, text length, spatial organization, and amount of white space all affect the visual structure of the document. Designers may want to base the visual structure of an electronic manual on the hard copy version of that manual, but the two do not have to be identical. In fact, some aspects of the hard-copy visual structure may need to be modified for electronic displays.

Equipment Recommendation(s)

> If frames are used, designers should display similar types of information in the same spatial frame.
>
> Use white space to separate sections of text.
>
> Where possible, data should be formatted into short segments, each of which communicates one clear point.

Problem Statement

Electronic documents need to be designed for ease of reading. Readability is affected not only by how legible the individual characters or words are, but also by how well the visual structure matches and reinforces the logical structure of the concepts in the material.

Example(s)

Short text segments are easier to comprehend than long text segments. Long segments of plain text are difficult to read, even if the font is well chosen. This visual structure should match the logical structure of the document in that each short text segment should be focused on communicating one clear point. Ideally, each text segment would be visible in its entirety within the display area of the EFB, so that the user does not have to access off-screen text to comprehend the point.

Use of frames can improve readability if frames are used consistently. For example, consider the case where the screen is divided into three frames. The upper half of the screen is one frame, and the lower half is divided into two equal sized frames. The top frame could be used to always display the main text of the document. The lower left frame could display a list of related links for navigation, and the lower right frame could always show related figures and/or tables (i.e., detail information).

Header and footer information that contains a brief description of the section heading can be useful for orientation.

Portable Document Format (PDF) documents, which capture the visual structure of the printed page, may need to be optimized for electronic presentation rather than just copying the paper version. For example, font choices may be modified.

Evaluation Questions

- Does the visual structure of the electronic document match and reinforce the logical structure of the document?
- Are frames (if available) used consistently?
- Is white space used to separate sections of text?
- Are text segments generally short?

3.3.2 Minimum Display Area

Equipment Recommendation(s)

> The manufacturer of an electronic document application should identify and specify the minimum display area necessary to view documents.
>
> Operators should meet the minimum display area requirements suggested by the manufacturer for both training and operational use of the electronic document application.

Problem Statement

In constructing the visual structure of an electronic document, the designer will have to assume a minimum display area. This assumed size will determine the smallest available space for structuring the data. Designers will make choices about the visual structure (e.g., font size and white space) based on this minimum display area. While the visual structure may transfer to a larger screen area without difficulty, it is not likely to work properly on a smaller display area that violates design assumptions.

Example(s)

If an operator intends to use the same electronic document software on EFBs with different display areas, they should ensure that all the different EFBs meet the minimum display area requirements suggested by the manufacturer of the electronic document application.

Evaluation Question(s)

- Does the manufacturer of an electronic document application specify a minimum display area?
- Does the EFB display intended for this application meet the minimum display area suggestions?

3.3.3 Off-Screen Text

Equipment Requirement(s)

> If the document segment is not visible in its entirety in the available display area, the existence of off-screen content must be clearly indicated in a consistent way.

Equipment Recommendation(s)

> If part of the document segment is off-screen, the following information should be constantly be available to the user:
> - How long the document segment is
> - How far in the document segment the currently displayed information is

Problem Statement

Documents that will be on an EFB are lengthy and complex. However, they can be broken down into natural segments, such as different sections and subsections. Still, it is possible that a single document segment will be too long to be displayed in its entirety within the available display area. Users then must be made aware of the existence of off-screen content. Plus, the software should be designed to assist users in managing what text is in view and what text is out of view.

With a paper document, the pilot can look ahead to check its length and they can browse or scan the entire document to orient themselves. On an electronic display, it is useful to convey the length of the document, either graphically or numerically. One way of helping the user to orient him/herself is to convey how far in the document the currently displayed information it.

Example(s)

If the document is implemented in terms of discrete "pages," then the current page and the total number of pages can be indicated using a convention such as "1/3," where the first number is the current page, and the last number is the total number of pages. Arrow buttons can also be used to indicate whether there are more pages preceding or following the page currently in view.

If the document is implemented on a scrolling window, a side scroll bar can convey all the required information. A graphical box or bar would represent the location of the currently displayed text in relation to the length of the document.

More sophisticated displays for user orientation, such as an outline view, can also be very helpful for user orientation.

Evaluation Question(s)

- Is the existence of off-screen text indicated clearly? Is the existence of off-screen text indicated in a consistent way?
- Does the software indicate how long the current document segment is and the position of the currently displayed information relative to the entire length of the segment?

3.3.4 Active Regions

Equipment Requirement(s)

> Active regions must be clearly visually highlighted.
>
> Note: Active regions are regions to which special user commands will apply. The active region could be text, a graphic image, a window, frame, or other document object.

Training/Procedures Requirement(s)

> Users must know how to activate and deactivate regions.

Training/Procedures Recommendation(s)

> Users should know the basic special commands are available for different types of active regions.

Problem Statement

It is often necessary to specify an active region to which special commands will be applied. For example, a text string might be selected for copying into a search query, or a window might be activated in order to bring it to the front of other windows on the screen. Active regions are also useful for selecting between frames on a frame-based visual display. The information in the active frame would respond to update commands entered by the user.

While active regions are not a required feature of electronic documents, if they are supported they must be clearly indicated and users must know how to use them. If the user does not know how to use an active region, he/she will have trouble applying special commands to the intended object. If the user does not know that a particular region is active, he/she may enter inappropriate commands and become frustrated when these commands are not processed as expected.

Example(s)

Active text could be highlighted using reverse video. Some special commands that might apply to active text include copying and deleting. Note that a sophisticated input mechanism, such as pen input, is required for text selection.

Active windows and frames could be highlighted with special borders. Once highlighted, the window would appear in the forefront (if multiple windows are supported). Only the active window would respond to scrolling, repositioning, or resizing commands. A highlighted frame could respond to scrolling commands, but might not be movable, or resizable.

Evaluation Question(s)
- Are all types of active regions clearly highlighted?
- Are users trained in how to activate and deactivate regions?

3.3.5 Display of High Priority Information

Equipment Issue

> Some parts of an electronic document may contain high priority information that might be accessed during critical phases of flight. Because the consequences of user error may be more significant under these conditions, designers may want to apply more generous standards for legibility and readability to high priority information, such as a larger typeface and more spacing between lines.

Problem Statement

Although electronic documents will primarily be used during low-workload conditions, it is possible that some parts can be expected to be used during critical phases of flight, which hold a greater potential for human error. The contents of electronic documents could be reviewed to identify those sections which are more likely to be accessed under conditions that increase the potential for error, the occurrence of which may produce more significant consequences. If more generous standards for legibility and readability are applied to high priority information, the potential for error could be reduced.

Example(s)

The limitations section, which may be found in the Flight Standards Manual, is one example of a section that might benefit from the application of visual structure that reduces the amount of information displayed at one time. A larger typeface, more spacing between lines, and the use of additional white space between chunks of information could reduce the risk of misreading key information.

There may also be more sophisticated ways of addressing this issue. For example, an electronic document application that is integrated with other cockpit systems could dynamically highlight relevant information on the document.

Evaluation Question(s)

- Are more generous legibility and readability standards applied to high priority information in the electronic document?

3.3.6 Figures

Equipment Requirement(s)

> At a minimum, the electronic version of a figure must be able to display all of the content of the paper version. The user must be able to view the overall figure at one time, even if not all the details are readable in order to get an overview of the figure. The user must also be able to read all the details in the figure, although not all of the figure may be visible when the details are readable.

Equipment Recommendation(s)

> Depending upon the size and complexity of the figure, and the available display area and resolution, the user may have to manipulate the figure to (a) bring areas of the figure that are out of view into view, or (b) make readable details of the figure that are not readable otherwise.
>
> The additional workload of manipulating figures is undesirable, so figures should be displayed in their entirety with all details readable whenever possible.
>
> Each figure should have descriptive text information associated with it. This text should be available even if the figure is not displayed.

Equipment Issue

> Figures could be designed to take advantage of the electronic medium in many ways. However, more flexibility in the manipulation of figures increases the complexity of using the software, which in turn impacts user training needs. Also, more flexibility in the manipulation of figures may actually distract users from understanding the actual content in the figure.

Problem Statement

Figures can be used for various purposes. Graphs are a type of figure. They illustrate relationships between variables. Graphs can also be used as a source for data for calculations. Other figures are representations of relationships between components of systems.

In paper form, figures are drawn so as to be usable for one or more purposes. The user may be interested in either the detail information in the graphic, the overall schematic information in the graphic, or both. The electronic version of the figure must capture all the content of the paper version, which includes both the details and the overall schematic information. However, there may also be additional workload and training needs that arise from the need to configuring electronic figures for maximum usability. Designers should consider these tradeoffs when implementing more complex features for electronic figures.

Example(s)

In order to load data more quickly, the figure may not be displayed until specifically called up. The associated text information should describe the contents of the figure so that the user knows whether they want to call up the actual figure.

Some ways in which figures could take advantage of the electronic medium include interactivity and customization. For example, figures could be interactive in that the user could specify which "layers" of the drawing they wanted to be visible or not, in order to understand complex relationships. A graph could also be interactive in that the user could select parameters for producing the plot.

A electronic document application that was integrated with other cockpit systems could also produce custom graphs or figures relevant to the current situation. For example, if the weight of the aircraft is known, graphs and figures could show data that apply for that aircraft configuration.

Evaluation Question(s)

- Can the user view the entire figure at one time? Can the user configure the display such that details of the figure are readable?

3.3.7 Tables

Equipment Issue

> Because of screen resolution limitations, translating tables from paper to electronic format may require modifications to their design to ensure that users can quickly locate target information contained in the tables.

Problem Statement

Tables typically provide a lot of data in a relatively compressed space. Borders and lines are often used to help the reader visually organize the information correctly. The comparatively low resolution of EFB screens, as compared with paper, may require that the design of existing tables be reconsidered to ensure that comparable readability is achieved.

Table size may also need to be reconsidered. Paper tables may utilize the full height and width of a standard page size. EFB screen size may be significantly smaller. A larger type size may also be required, which may require rethinking of how much data can be supported by an individual table.

Example(s)

Additional white space may be required to clearly separate individual table elements from each other and also from neighboring borders or lines. Placing column and row names in bold can help the user to interpret table information more efficiently.

If the electronic document application is aware of the aircraft systems' status, it may be possible to customize and reduce the information in a table such that only the relevant data is displayed. For example, if the weight of the aircraft is known, or the ambient temperature is known, data that apply only to those and similar conditions could be displayed.

Evaluation Question(s)

- Are the tables included in the electronic documents as readable and as usable as their paper counterparts?

3.4 Navigation and Searching

3.4.1 Moving to Specific Locations

Equipment Recommendation(s)

> If the electronic document application supports links, entries in the document table of contents and indices should be linked to the corresponding locations in the text. Cross-references should also be linked to each other, both within and across documents.
>
> If the electronic document application supports user customization, users should be able to configure and manage their own bookmarks to selected locations in the text.
>
> The electronic document application should keep track of the most recently visited locations in the document and allow the user to select from this list to return quickly to a recent location.
>
> The user should be able to cancel a movement by returning to previous location in one step.

Problem Statement

The manuals that pilots use are typically lengthy documents. It is important that users be able to navigate quickly to important locations within these lengthy manuals. Key locations in the document include the beginnings of each new section, which are listed in the table of contents, cross-references, and index entries. Each of these types of information can and should be linked to the appropriate location in the text. Other key locations in the manuals include recently visited locations and entries in the indices.

Because users often find themselves revisiting certain parts of an electronic document, it is also desirable to implement customizable electronic "bookmarks." These bookmarks would be set and managed (e.g., renamed or deleted) by the user.

Sometimes users will make a mistake and unintentionally move to a location. For these cases, the user should be allowed to cancel the movement and return to their previous location in one step.

Example(s)

Hyperlinks to different locations in the text are one way to move about a lengthy document quickly. Another way of moving about the document could be by selecting a location to move to from a list. For example, the last five visited locations could all be listed in one place (a "catalog" page), and the user could select which of these locations to move to by selecting one from the list. The selection need not be done by a pointing device. Selection could be accomplished through soft keys, which change the selected location.

Evaluation Question(s)

- Are the table of contents and indices linked to the corresponding locations in the text?
- Can users customize their own bookmarks to specified locations in the text?
- Can the users quickly return to recently visited locations in the text?

3.4.2 Managing Multiple Open Documents

Equipment Requirement(s)

> If the electronic document application supports multiple open documents, the title for the active document must be displayed continuously. Also, the user must be able to choose which of the open documents is currently active.
>
> Note: The active document is the one that is currently displayed and responds to user actions.

Equipment Recommendation(s)

> If the electronic document application supports multiple open documents, a master list of all open documents should be available.

Equipment Good Practice(s)

> Access to a document can be provided through the master list of open documents.
>
> If the display area is large enough, it is useful to be able to arrange multiple open documents such that text from more than one document is visible at the same time.

Problem Statement

If the electronic document application can support multiple open documents, the user will need assistance in managing these documents, or else they may become confused about what document they are using because the visual structures of the documents may be very similar. In particular, the user must keep track of which documents are open, and which of these is the currently active document. Document titles help the user manage multiple open documents. The user must also be able to move between the open documents quickly. That is, the user should be able to activate an open document from a list of open documents, rather than from a list of all available documents

Example(s)

If the electronic documents are running on an industry standard windows-based GUI, document titles should appear at the top of the window frame by convention. The different documents would be selected by activating the particular window that contains the document of interest.

The user may want to open both the POH and the MEL in order to see some cross-referenced items. He/she may want to arrange these two documents such that the top half of the screen shows one document, and the bottom half shows text from the other.

Evaluation Question(s)

- If multiple open documents are supported, is the title of the active document shown continuously? Can the user easily choose which open document is active?
- Is a master list of open documents available?

3.4.3 Searches

Equipment Recommendation(s)

> The electronic document application should support multiple search techniques. Some options include searching by:
> - key word
> - links to text (e.g., via cross-references or a table of contents)
> - graphical links (e.g., look up the function of a switch based on its location in the cockpit)
> - header/footer information (i.e., brief topic information).

Equipment Issue(s)

> If key word search is implemented, the EFB must support entry of free text, preferably via key board rather than soft keys. Also, designers will have to consider the design of a query language. Complex query languages may have training implications. Also, designers will have to consider how the user will move between multiple hits from a key word search.

Problem Statement

One of the advantages that electronic documents can have over paper documents is the ability to search a lengthy document quickly. With paper documents, search is conducted through section headings, or indices. There are other options that can be implemented electronically, such as key word search, and text and graphical links. These electronic search techniques can be very effective when the user is conducted a very well specified search. Browsing header/footer information can be very effective when the object of the user's search is specified well enough for other types of searches.

Example(s)

To find definitions for various levels of turbulence, the user could do a search on the keywords "turbulence" and "definition," or he/she could search via links from the table of contents of the flight operations manual.

Evaluation Question(s)

- Are multiple search techniques supported in the electronic document application?

3.5 Options

3.5.1 Links to Related Material

Equipment Good Practice(s)

> Access to related information (e.g., definitions of acronyms and terms) could be provided within the electronic documents through links, pop-up information, or other similar techniques. The user could also have easy access to information on related topics and complementary information such as checklists and the MEL.

Equipment Recommendation(s)

> If related information is accessible within the electronic document, a consistent philosophy should be used for determining how different types of information will be accessed. Similar types of information should be accessed in the same way.

Problem Statement

While pilots are very familiar with the reference manuals they use, it is desirable to increase the usability of these documents by building in links to related information. If links are available, they may allow the pilot to obtain all the necessary information from one source, rather than requiring him/her to open multiple documents or document segments to obtain the information. Links are also helpful when the pilot does not have a focused question in mind, but rather is trying to obtain more general information.

The type of information being related should determine the access method used. A consistent access philosophy will help users to anticipate what they will find if they access linked information.

Example(s)

Examples of access methods include hyperlinks to other sections of a document, pop-up panels that provide *brief* definitions of terms and acronyms (e.g., "balloon help"), context-sensitive help, and navigation buttons or tabs displayed on the screen.

Definitions of words and acronyms can be provided through pop-up panels. The pop-up panels could be activated when a word or acronym is selected and the "help" feature is on, or they could be identified by a special character formatting and then accessed when the term is "clicked."

Hyperlinks are useful to enable the user to access additional information on a referenced topic. They are especially appropriate for supporting navigation from one location in a document to another document section that provides more detail.

Navigation buttons or tabs are useful for supporting access to complementary types of information. For example, the operations manual, which dedicates a chapter to each aircraft system, could display navigation tabs at the bottom of the screen which support access to checklists and MEL information from each system chapter.

If a consistent philosophy of access to information is implemented, all definitions could be accessed in one way (e.g., pop-up panels), and all links to other document segments could be implemented in a different way (e.g., navigation buttons). The user then expect different types of information when he/she clicks on a terms as opposed to when they click on a navigation button.

Evaluation Question(s)

- Is access to related information supported by the electronic documents?
- Are similar types of information accessed in the same way?

3.5.2 Display Customization

Equipment Issue

> The extent to which the information display can be customized must be carefully evaluated to ensure that degradations in usability, legibility and readability do not occur. It may be appropriate to limit the types and range of customization for use on the flight deck relative to that which is typically provided for a standard graphical user interface.

Equipment Requirement(s)

> If the electronic document application supports customization, it must also provide an easy means by which to reset all customized parameters back to their default values.

Problem Statement

Users may want to be able to customize the appearance of the electronic document. Certain customization features can be helpful, such as allowing the user to increase the font size. However, customization may actually result in degraded information appearance. Therefore, it may be necessary to limit what parameters may be manipulated and the range of manipulation supported.

An EFB that is installed on the flight deck will be used by multiple users, so there must be a simple command to restore the appearance of the electronic document to its default parameters. In addition, providing an easy means for restoring the default parameters will reduce the time spent re-customizing the display and minimize the potential distractions associated with customization.

Example(s)

Personal computers allow users to manipulate resolution, background and typeface colors, font and document size, and a host of other parameters. The flight deck environment may not be the place to experiment with this broad flexibility. Each customizable parameter should be assessed as to its potential for reducing the readability and legibility of the information presented.

Unlike most office computers, EFBs may often be used by more than one user. Customizable parameters could be stored centrally, thus reducing the amount of time spent manipulating the parameters in the first place and allowing for a single reset button that would return all parameters to their default settings.

Evaluation Question(s)

- Does the electronic document application provide an easy means for resetting all parameters to their default values?
- Can the manipulation of a display parameter produce a significant decrement in the appearance of the displayed information?

3.5.3 Printing

Equipment Recommendation(s)

> Users should be able to select the subset of information they wish to print, including individual sections and individual pages.
>
> The visual structure of the printed document should match the visual structure of the electronic document.

Equipment Requirement(s)

> The electronic document application must clearly specify which pages or document sections have been selected for printing.
>
> The user must be able to immediately terminate a printing session if a large printing range has been selected.

Problem Statement

Users may wish to have a hard copy of some portion of an electronic document. If available, the printing option should enable users to select, as precisely as possible, the subset of information of interest so that the user is not inundated with irrelevant material. In addition, users must be able to terminate a long print job in the event that an error has been made, so that the user can quickly return to the application to print the correct section.

Example(s)

The printing option should allow users to choose between printing either a page range (including a single page) or one or more sections. The print window must clearly indicate the print range that has been selected. To help ensure that the correct pages have been selected for printing, the printed version should closely correspond to the electronic version in terms of the subset of information that is printed is identical to the information displayed electronically.

In the event that an incorrect page range is selected, the user must be able to immediately terminate the printing session if more than a few pages have been selected for printing. Otherwise, the printer will be temporarily unavailable for further use and paper will be wasted.

Evaluation Question(s)

- Does the EFB allow users to choose the subset of information to be printed?
- Does the printed version correspond to the electronic version?
- Does the EFB clearly specify the subset of information that has been selected for printing?
- Can the user immediately terminate a print job that is larger than a few pages?

3.5.4 Animation

Equipment Requirement(s)

> If animation is supported, the user must be able to start and stop the segment. The user must be able to stop the animation at any time, even if the segment has not ended.
>
> Also, there must be supporting text to describe and support the animation. This text must be available even if the animation is not currently running.

Equipment Recommendation(s)

> Animation should only be used to highlight and explain important relationships. It should not be overused.
>
> If the animation has associated supplemental audio, control of both the audio and video should be integrated.

Problem Statement

Animation can be a powerful aid to visualization of complex relationships. It is especially useful for training or study of new, or very detailed material. In these situations, the user must focus attention on the animation for it to be of value.

Because of the need to focus attention on complex animations, its use in flight may need to be limited. At the very least, the user must be able to interrupt any animation in progress, in order to quickly change to higher priority tasks. Also, supporting text must be available to preview the content of the animation. In other words, the user must not be required to start the animation in order to determine its contents.

While animation can be beneficial for specific purposes, overuse of animation should be discouraged. Too much visual movement on the screen can distract the crew from higher priority tasks.

Example(s)

Animation could be implemented to show how parts of a complex mechanical system fit together, or as part of a training tutorial on use of the software.

Animation should not be used to highlight company logos.

Evaluation Question(s)

- Can the user control when the animation begins? Is the animation interruptible?
- Is there supporting text for the animation that identifies its contents without running the segment?

3.5.5 Making Notes

Equipment Issue(s)

> Paper documents are customizable in that users can make notes and highlight information of interest to them. Adding these customization features to the electronic document functionality could also be useful. The utility of this feature may be constrained, however, if users do not have access to their own notes while using an EFB. To ensure that the users can view their own notes, they may have to enter the notes on a personal EFB, or their notes may have to be accessible through a server that communicates with all EFBs.

Problem Statement

One of the advantages of paper documents is the ability for the user to underline, scribble notes in the margins, and otherwise utilize the document to record their own information. Users would benefit from having a similar capability with the electronic document function. However, the EFB may not be a personal unit, in which case, the notes may not be very useful to the other users.

The extent to which a notes feature would be useful will be determined in part by the company policy on EFB use. EFBs that are used only on the flight deck may benefit less from this feature in that users would have access to these notes only while on that specific aircraft.

Example(s)

The electronic document function could provide the capability to select chunks of text that could be highlighted in a fashion similar to underlining in a paper document. Similarly, a location in the text could be selected and a notes feature utilized to allow the user to record their own notes. Retrieval of notes could take place in two ways. First, access to the set of all notes that have been recorded could be supported. Second, a visual indicator could be placed in the location from which the note was created which would then be used to access the individual note that was created for that location. It is possible that these notes could be stored in a central ground-based server that would allow upload to the specific EFB that will be used by the note creator for each flight. If the server is used to store other information "owned" by that user, this might be useful to do.

Evaluation Question(s)

- If the electronic documentation supports note taking, can users always access their personal notes?

3.5.6 Decision Aid/Automatic call-up of Data

Equipment Issue

> The integration of an EFB with other flight deck systems may enable electronic documents to serve as a decision aid. Consideration should be given to integrating electronic documents into closed-loop systems to provide users with immediate access to information that can support more effective flight management. Although decision aids can have value in supporting more effective decision making, they can also have the unintended consequence of excessive reliance, where the crew may become complacent about reviewing the suggestions made by the system.

Problem Statement

Integrating the EFB with other flight deck systems could allow the electronic documentation application to customize its information based upon current flight conditions. Doing so can help to reduce crew workload by reducing the amount of information the crew must consider. An unintended consequence can be complacency where the crew relies on the decision aid to select information for review without sufficient crew involvement.

Example(s)

"Awareness" of aircraft system status could enable the electronic document application to display only the subset of data in a table that applies to the current situation. For example, knowing the fuel burn rate would enable the decision aid to list only the candidate airports within range if the crew needed to divert to a different airport.

Several approaches can be taken to mitigate problem of complacency. First, the decision aid could offer several options rather than a single answer; doing so would encourage the crew to review the information needed to select an option. A second approach is to require the user to make the decision first, then have the decision aid review the soundness of that decision.

Evaluation Question(s)

- Does the decision-aiding function mitigate the risks of crew complacency?

4 Electronic Checklists

Electronic checklists (ECLs) are a logical application for EFBs as well. In Section 4.1 below, electronic checklist concepts and terms are explained, as well as many of the potential benefits that they offer. Guidance statements for ECLs are presented in Sections 4.2, 4.3 and 4.4. Section 4.2 covers how checklists are accessed. Section 4.3 covers how individual checklists are completed. Optional ECL features are discussed in Section 4.4. All of these issues are important from an Equipment perspective, and some have implications for Training/Procedures.

4.1 Background

Effective use of checklists is a critical component of the pilot's job. Failure to complete a checklist has been implicated as a key contributor to several major accidents, reinforcing the need to present electronic checklists in such a way that pilots are encouraged to perform and complete them in a timely fashion.

Electronic checklists (ECLs) are a relatively recent addition to the flight deck environment. The benefits they provide depend upon the extent to which they utilize the advantages of the electronic medium. ECLs can simplify the process of accessing desired checklists. The simplest ECLs may be no more than a digitized, static version of paper checklists but ECLs with increased functionality will be much more desirable to users.

There are three significant features that can greatly enhance ECL functionality. First, the ECL could support sequencing of checklists. These sequences can be of three types:

- Implementation of the normal sequence of use. Because normal checklists are performed in a fixed order, the EFB can be programmed to display checklists in the prescribed order. This capability helps to remind the crew of the next checklist they need to accomplish.

- Implementation of embedded checklists which are accessed from a "parent" checklist. The user could break away from the parent checklist and move to an embedded ("child") checklist which, upon its completion, returns the user to the parent checklist.

- Access to related checklists on a need basis. Some non-normal conditions require the performance of multiple checklists. For example, a non-normal condition such as a power plant failure may require that a second checklist be used to accomplish a single-engine landing. Direct access to the second checklist can be supported from the first checklist. This capability helps to lead the crew through all of the checklists they need to complete, which offers significant task management benefits.

Coupling the ECL with aircraft systems can introduce additional benefits that reduce the crew's workload. During the performance of normal checklists, a closed-loop ECL can monitor crew performance of some items. Detection of a non-normal condition enables the ECL system to produce the required checklist at the crew's request. Considerations that address checklist access are described in Section 4.2.

A second useful feature of advanced ECLs would be an ability to indicate the status of individual checklist items and the checklist as a whole. In addition to being either active or inactive, a checklist item could be uncompleted, deferred, or overridden. A deferred item is one that the crew has skipped but intends to complete at a later time. An overridden item is one that the crew does not intend to complete. Finally, for checklist items presented by a closed-loop system, a checklist item can be either sensed or not sensed. A sensed item means that the ECL is able to detect when the crew has correctly completed the item.

An ECL that keeps track of item status can support additional functionality, such as indicating the completion status of the checklist as a whole in order to prevent the crew from closing an uncompleted checklist. Checklist item status and checklist completion functionality offer significant value in protecting crews from the deleterious effects of distraction. Considerations that address checklist and checklist item status management are described in Section 4.3.

Finally, ECLs are capable of providing additional functionality not available with paper checklists. For example, they may incorporate access to supplemental information such as clarification of the meaning of a checklist item or linked calculation worksheets. Also, ECLs can provide additional support for task management through the implementation of task reminders, which remind the crew to finish tasks that take significant time to complete. Considerations that address the use of these optional features are discussed in Section 4.4.

Version 1 (9/28/00)

4.2 Call-up/Access

4.2.1 Checklist Scope

Equipment Requirement(s)

> All checklists that belong to a category that is supported by the ECL must be available on the ECL.
>
> If an ECL-supported checklist requires access to a checklist that is not supported, the ECL-supported checklist must indicate the location of the unsupported checklist in the paper document.
>
> Note: Checklist categories typically include: normal, abnormal, emergency, and supplemental.

Training/Procedure Requirement(s)

> Flight crews must be trained to know which checklist categories the ECL supports.

Problem Statement

An ECL that does not include all of the checklists that belong to an ECL-supported category can create confusion and lack of trust in the system when crew members attempt to find a checklist that is not in the ECL. Excluding an entire category of checklists should not create confusion, such as using paper versions of the QRH together with electronic versions of the normal checklists.

Example(s)

Providing all of the checklists in a supported category removes the possibility that crews will waste valuable time attempting to find an unsupported checklist. If a supported checklist requires subsequent use of an unsupported checklist, the location of the unsupported checklist must be provided in the electronic checklist. Listing the paper location replaces a hyperlink or other electronic support that would otherwise be provided if both checklists were accessed from the ECL.

Evaluation Questions

- Does the ECL include all checklists for each supported checklist category?
- Is the location of the paper checklist provided in all electronic checklists that require subsequent access to an unsupported checklist?
- Is training provided that addresses which checklists are supported?

4.2.2 Accessing Normal Checklists

Equipment Requirement(s)

> If the ECL supports normal checklists, they must be accessible in accordance with the normal sequence of use during line operations.
>
> Normal checklists must also be individually accessible at all times.

Problem Statement

Normal checklists will be accessed during line operations, and may also be accessed for review. Their primary use is to ensure proper aircraft configuration at critical points during each phase of flight, including pre- and post-flight. ECLs must present checklists in the normal sequence to prevent the crew from skipping a checklist by mistake.

Users must also be able to access individual checklists at any time should they desire to review a specific checklist.

Example(s)

Normal checklists are typically performed in a fixed order, so the ECL may be designed to call up the next checklist automatically after the completion of its predecessor. With this feature, the crew would not be required to locate and access the next checklist.

For closed-loop systems, all checklists could be automatically displayed in response to the occurrence of appropriate system-based triggering conditions. A tie-in to aircraft systems means that system changes could determine which checklist to display at any given time.

Access to individual checklists could be supported from a "table of contents," which could be a menu or list of checklist titles that are hyperlinked to the actual checklists. The organization of the contents list should be consistent with that used in the FAA-approved flight manuals.

Evaluation Questions

- Are normal checklists presented in order during line operations?
- Can each normal checklist be individually accessed easily?

4.2.3 Accessing Non-normal Checklists

Equipment Requirement(s)

> If the ECL supports non-normal checklists, access to individual non-normal checklists must be supported at all times.
>
> When a non-normal condition is detected by a closed-loop ECL during line operations, the ECL must alert the crew that a checklist applies to this condition. The ECL must only call up the appropriate checklist when commanded by the crew.

Problem Statement

Like normal checklists, non-normal checklists will be accessed for several purposes. Their primary use is to support management of a non-normal condition. During the high workload conditions that often accompany management of a non-normal condition, the potential for selecting the wrong checklist is high. Accurate, rapid access to the appropriate non-normal checklist is required. In addition, pilots will want the ability to access individual checklists at any time for review.

Example(s)

Non-normal checklists are used on a need basis. A specific checklist could be accessed manually by means of the table of contents. Organizing checklists by the type of non-normal condition, as found in Quick Reference Handbooks, may be appropriate. Providing clearly visible checklist titles, together with a list of the indicators for the corresponding non-normal condition, can help crews to avoid selecting an incorrect checklist. This is particularly important for non-closed-loop systems.

In addition to manual access to individual checklists, closed-loop systems also support automatic access in response to system detection of a non-normal condition. An alert occurs and access to the appropriate checklist is provided when commanded by the crew.

Evaluation Questions

- Can each non-normal checklist be easily accessed?
- Is the appropriate non-normal checklist easily accessed when a non-normal condition is detected by a closed-loop ECL?

4.2.4 Open Checklists

Equipment Requirement(s)

> If the ECL supports more than one open checklist, the user must be able to access other checklists without having to close the currently displayed checklist first.
>
> Each checklist must have a constantly visible title to ensure that the user always knows which checklist is currently displayed.
>
> Note: An open checklist has been accessed by the user and is tracked by the ECL as "open." This open status means that if an open checklist is out of view, it can be accessed through quick operations that apply only to open checklists. For example, an ECL might provide a toggle button for moving only between open checklists or a master list that only includes open checklists. A closed checklist no longer receives this special tracking and cannot be accessed through operations that apply only to checklists having the open status.

Problem Statement

To manage some flight conditions effectively, the crew may want to access more than one checklist at a time. Flexible use of multiple checklists requires that the crew be able to move quickly between open checklists. Requiring the crew to close one checklist in order to open another checklist is inefficient and may discourage crews from fully utilizing checklists.

Example(s)

The crew is performing the 10,000 foot/climbing checklist. A left cowl anti-ice message appears on the EICAS. Because the cowl anti-ice is on, a low pressure condition is indicated. The climb checklist is not complete but they need to open the cowl anti-ice checklist because they are in icing conditions. They perform the first part of the checklist which ends with the command to leave icing conditions. After reconfiguring the aircraft to depart the icing conditions, they return to the climb checklist. Then they return to the cowl anti-ice checklist which, on the EFB, includes a natural transition to the ice dispersal procedure. Both of the open checklists are titled so that the crew knows which one is in view at any time.

Evaluation Questions

- Can the user easily transition to other open checklists without closing the current checklist?
- Does each checklist have a constantly visible title?

4.2.5 Multiple Open Checklists

Equipment Requirement(s)

> If more than one unrelated checklist can be open, the user must be able to choose which checklist is currently active.
>
> Note: The active checklist is the one which is currently displayed and the status of its items change in response to user actions.
>
> If one checklist is a "child" of another checklist in which it is embedded, the user must be able to choose whether the parent or the child checklist is active.
>
> Note: If one checklist is embedded in another, the higher-level checklist is called the "parent," and the embedded (lower-level) checklist is called the "child."

Equipment Recommendation(s)

> In place of parent-child checklists, create a single checklist that incorporates both.

Problem Statement

If multiple checklists can be open at the same time, the user must be able to choose which one is he/she is actively working on so that he/she can prioritize the order in which the checklists will be performed or so that he/she can review other checklists without closing the checklist in progress. If the user can not select which checklist is active, then the last checklist that was opened will be the default checklist in progress, but it may not be the highest priority one.

Examples

The user should be able to move between unrelated checklists, as well as related checklists. For example, a checklist with an embedded child checklist must allow the user to access child checklist. Once the child checklist is active, there must be a way for the user to return to the parent checklist.

Evaluation Questions
- Can the user move from one open checklist to another easily?
- Can the user move from a parent checklist to a child checklist easily?

4.2.6 Managing Multiple Checklists

Equipment Recommendation(s)

> If more than one checklist can be open at one time, a master list of all open checklists should be provided.

Equipment Good Practice(s)

> Access to an open checklist can be provided through the master list.

Problem Statement

Keeping track of multiple open checklists can be challenging when other tasks compete for the crew's attention. By providing a list of all open checklists, it is easier for the crew to be aware of pending checklist tasks.

Examples

A master list of all open checklists should be provided that is easily accessed at all times. This checklist queue can also be used to cycle through all open checklists.

Evaluation Questions

- Are open checklists all listed in one place?

4.2.7 Managing Multiple Non-Normal Conditions

Equipment Requirement(s)

> For closed-loop systems, all checklists required to manage multiple non-normal conditions must be listed together in one master list.

Equipment Recommendation(s)

> The master list of checklists to be completed for managing a non-normal condition should indicate the status of each checklist (e.g., pending, or completed).

Equipment Good Practice(s)

> Access to a specific checklist during a non-normal condition can be supported from the master list of required checklists for that condition.

Problem Statement

Multiple non-normal conditions may require the use of more than one checklist. Keeping track of which checklists need to be performed and the status of each checklist places additional demands on a crew that may already be overloaded. A closed-loop ECL can help the crew manage the performance of all required checklists while providing crews the flexibility to schedule checklist performance in accordance with other high-priority tasks.

Examples

A list of all checklists that must be completed for multiple malfunctions must be provided. The status of each checklist could be indicated within the master list. For systems such as the EICAS, the warning messages can indicate whether corresponding checklists exist. The indicator can also change color or otherwise reflect the status of its checklist.

In addition, this master list could provide direct access to the associated checklist. Doing so minimizes the potential for selection of the wrong checklist. A checklist button could be used to cycle through all checklists that must be completed, including those which were open when the non-normal conditions occurred.

Evaluation Questions

- Does the ECL indicate the checklists that must be performed when multiple malfunctions have occurred?
- Can the user easily access these checklists?

4.2.8 Putting Away the Checklist

Equipment Recommendation(s)

> The ECL should allow a state where there are no currently open checklists.

Problem Statement

Checklists are not in use for a large proportion of a routine flight. With paper checklists, the crew simply puts the checklist back into storage. The act of storing a paper checklist indicates that the checklist has been completed and removes the checklist as a distraction. An equivalent capability should be supported by any ECL. Although an ECL may keep track of the next checklist to be used, it should not automatically open that checklist upon completion of the previous checklist.

Example(s)

After completing the take-off checklist and passing beyond 18000 feet in US airspace, there are no checklists to use under routine conditions. The checklists are stowed. On an ECL, a blank screen would indicate that there are no currently open checklists, as would a text message to that effect.

Evaluation Questions

- Does the ECL allow a state where there are no currently open checklists?

4.3 Checklist Actions

4.3.1 Indicating the Active Checklist Item

Equipment Requirement(s)

> The ECL must provide a pointer that indicates the active item in the checklist.
>
> If more than one unrelated checklist can be open or if the EFB supports multiple functions that can interrupt checklist completion, a placeholder capability is required to remind the user which item was active prior to leaving the checklist.

Problem Statement

A number of accidents and incidents have occurred, at least in part, because of the failure of the crew to complete all of the items in a checklist. Distraction, high workload, and other factors may cause the crew to unintentionally skip an item. Similarly, moving between checklists or other EFB functionality can result in the user losing his/her place in the checklist. Even with systems that indicate individual item status, additional time is required to identify the first uncompleted item in the list.

Example(s)

A pointer or box that surrounds an item can be used to indicate the active item. When returning to a checklist that was started but not completed, the item that was active prior to the move should again be active.

Evaluation Questions
- Is the active item clearly indicated?
- Does the ECL indicate the item in the checklist where the user was prior to leaving the checklist?

4.3.2 Moving Between Items Within a Checklist

Equipment Requirement(s)

> Moving the active-item pointer to the next checklist item must require only a simple action by the user.
>
> For ECLs that track the status of individual checklist items, the user must be able to move backward through checklist items to return to a previous item without changing the status of any of the items.

Equipment Recommendation(s)

> For ECLs that track the status of individual checklist items, the user should be able to:
> - Move from an uncompleted checklist item to the next item in the checklist, changing the status of the uncompleted item to "deferred."
> - Move to the next item in the checklist automatically after a completing an item.

Equipment Issue(s)

> Designers should consider how quickly the user is allowed to move to another item. If rapid movements are allowed, the user may unintentionally skip beyond the desired item. If the user must stay on each item for a small period of time, the chances of moving beyond the desired item are reduced, but this solution may increase the overall amount of time required to perform the checklist.

Problem Statement

EFB users need to be able to move easily and quickly between items in a checklist. If users are not allowed to move around the checklists flexibly, they may take more time to complete the checklist, or they may become so frustrated with the ECL that they do not use it properly.

Example(s)

ECLs that use an active-item pointer must support easy movement to the next item in the checklist. This movement must require only a simple user action. For ECLs that track the status of individual checklist items, moving between items must not affect their status, except in the case of moving to the next checklist item from an uncompleted item. In this case, the status of the prior item should change from uncompleted to deferred.

In addition, the ability to move backward must be implemented to avoid forcing the user to move forward through all checklist items in order to return to the desired item. Backward movement alone must not change a checklist item's status.

Evaluation Questions

- Is it easy to move the active-item pointer to the next checklist item?
- Can the user move backward to a previous checklist item without affecting the status of any item? If the user moves forward in the checklist, are deferred items marked appropriately?
- Does the active item change to the next one in the list after an item is completed? Is there a tendency to skip items when attempting to move to the next item?

4.3.3

4.3.3 Specifying Completion of Item

Equipment Requirement(s)

> If the ECL requires a user action to indicate item completion, this action must be simple and distinct from the action of moving to the next item.

Equipment Recommendation(s)

> If the ECL requires an action to indicate item completion, the act of marking an item as complete should cause the next item in the checklist to automatically become active, except if the item is on the next page. A separate action should be required to move to the next page.
>
> Should an item have been incorrectly designated "complete," an easy undo should be available.

Problem Statement

An important advantage of electronic checklists is their ability to indicate which items within a checklist have been completed. This feature reduces the likelihood that an item will not be completed. An action separate from that of moving to the next item in the checklist should be required to change an item's status to "completed."

Example(s)

The user could press a button to indicate completion of an item. This button should be different from another button that might be used to move to the next item in the checklist without first completing the item (i.e., the other button would mark the task as "deferred" instead of completed).

Evaluation Questions
- Is the completion status of each checklist item indicated clearly?
- Does the action required to change an item's status differ from the action required to move to the next item if the item is not completed?
- When the status of an item has been changed to indicate completed, does the next checklist item automatically become active?
- Is a separate action required to move to the next page after all the items on the current page are completed or deferred?

4.3.4 Closing a Checklist

Equipment Requirement(s)

> If the ECL keeps track of item status and the user attempts to close an incomplete checklist, a warning must appear indicating that the checklist has not been completed.
>
> The user must be allowed to close an incomplete checklist after this warning has been acknowledged.

Problem Statement

One way to remind users to complete a checklist is to keep the checklist open until it has been completed. Once the checklist has been closed, the user may not remember to open it again and complete it without a reminder. There may, however, be circumstances where the user can reasonably want to close an incomplete checklist. Therefore, the user must always have the ability to close an incomplete checklist.

Example(s)

If the user attempts to close a checklist that has not been completed, a warning message could appear that reminds the user that the checklist has not been finished. The user would then be given the option to either go ahead and finish the checklist or to close it without completion.

Evaluation Questions
- When closing an incomplete checklist, is the user adequately warned?
- Is the user still allowed to close the checklist?

4.3.5 Undoing An Item Status Change

Equipment Requirement(s)

> If the ECL requires a user action to indicate item completion and the active item has been marked "complete," changing its status either to a different status or to return it to the uncompleted status must be simple to accomplish.

Problem Statement

Pilots may make errors while completing a checklist. The ECL must support error recovery by enabling easy modification of an item whose status has been incorrectly modified.

Example(s)

Changing the status of an item could be implemented by selecting the item to be modified and then selecting the "return to uncompleted status" button. After the item is back in the uncompleted state, the user may change its status again if necessary to get to a different state, e.g., deferred.

Evaluation Questions

- Is it easy to change the status of an item to a different status, including uncompleted?

4.3.6 Displaying Item Status

Equipment Requirement(s)

> If the ECL indicates the status of an item (active, deferred, overridden, uncompleted, closed-loop sensed), a clear visual indication of that status must be provided.

Problem Statement

Users can be aided by having the ECL indicate the status of each item on a checklist. In addition to being either active or inactive, a checklist item can also be uncompleted, deferred, or overridden. A deferred item is one that the crew has skipped but intends to complete at a later time. An overridden item is one that the crew does not intend to complete. Finally, for checklist items presented by a closed-loop system, a checklist item can be either sensed or not sensed. A sensed item means that the ECL is able to detect when the crew has correctly completed the item.

Example(s)

Each checklist item state that is supported by the ECL must have a unique visual code that can be quickly discriminated under a range of lighting conditions. The code that is used should be consistent with other color code applications in the flight deck. Deferred items can also be moved to the end of the checklist to indicate that they remain incomplete.

Evaluation Questions

- How are the possible checklist states indicated and are they easy to recognize?
- Are the visual codes sufficiently unique as to be clearly discriminable under all likely lighting conditions?

4.3.7 Returning to Deferred Items

Equipment Requirement(s)

> If the ECL keeps track of item status, then before a checklist can be declared complete, the user must be required to return to deferred checklist items and complete or override them.

Problem Statement

One of the advantages of an EFB is its ability to remind users to complete all items on a checklist. Items might be deferred unintentionally, due to distraction, intentionally, for workload management or because the conditions are not right to complete the item, or because an unexpected event requires a transition to a non-normal checklist. If these items were not addressed at some point, the checklist would not be complete. To ensure that the checklist is completed, an ECL that keeps track of item status must remind the user to complete or override deferred items.

Example(s)

Reminders could be implemented at two levels. First, there could be visual indicators within the body of the checklist that indicate the presence of deferred items. The item itself could be color coded as a deferred item or the item could be moved to the end of the checklist. In addition, the checklist itself could have a visual indicator showing that it has deferred items. Also, a code could be used in a master list of open checklist that reflects the presence of deferred items in an open checklist.

Evaluation Questions

- Does a master list indicate the presence of one or more deferred items in an open checklist?
- Can a checklist be closed if it contains one or more deferred items?

4.3.8 Integrating Non-Normal Items into Subsequent Checklists

Equipment Recommendation(s)

> Non-normal checklist items that are to be performed at a later time should be automatically integrated into subsequent checklists.

Problem Statement

Non-normal checklists can contain items that must be performed during subsequent phases of flight, in particular, approach and landing. Because they are to be performed at a later time, busy crews may forget to perform them at the appropriate time. Integrating these items into subsequent checklists can ensure that the tasks are performed at the correct time.

Example(s)

Instead of keeping incomplete checklists open, the ECL can move these items to a later checklist, allowing them to be performed at the appropriate time. An indication should be provided that these items originally came from a different checklist.

Evaluation Questions

- Does the ECL incorporate non-normal items into the appropriate checklists?

4.3.9 Lengthy Checklists

Equipment Requirement(s)

> A multi-screen checklist is one that has more items than can be displayed at one time on the EFB display. The ECL must allow the user to look ahead in a multi-screen checklist without changing the active item.
>
> If the user makes a change to an active item that is out of view, that active item must be brought into view.

Equipment Recommendation(s)

> While a multi-screen checklist is in use, the following information should constantly be available:
>
> - How long the whole checklist is
> - How far down the checklist the currently displayed information is
> - How much of the checklist has been completed.

Problem Statement

Some checklists are lengthy. With a paper checklist, the pilot can look ahead to see how long the list is, and judge how far along he/she is down the list. On an electronic display, the checklist may contain more items than fit on one screen at a time, even if the screen is relatively large. For task scheduling and other purposes, it is important to know how long the checklist is and how much remains to be completed.

While working on a lengthy checklist, the user must keep track of the current active item at all times. With a paper checklist, this can be done mentally, or with a physical reminder such as finger placement. With an ECL, the active item is tracked by the system, and must be displayed when a change is made to that item.

Example(s)

If the checklist is implemented in terms of discrete "pages," where each page represents the number of items that can be displayed at one time, then the current page and the total number of pages can be indicated using a convention such as "1/3," where the first number is the current page, and the last number is the total number of pages.

If the checklist is implemented on a scrolling window, a side scroll bar can convey all the required information. For example, if half the checklist has been accomplished, the graphical box or bar would be positioned midway down the vertical length of the window.

If the user moves ahead to view later pages or scrolls to a location where the active item is out of view, the active item must not change. If the active item is out of view and the user attempts to change the status of the item, the action must bring the active item into view.

Evaluation Questions

- When viewing looking ahead in a multi-screen checklist, does the active item remain unchanged?
- Is the active item brought into view when it is out of view and the user makes a change to it?

4.3.10 Confirming Completion of Checklist

Equipment Recommendation(s)

> The user should receive a positive indication that the checklist as a whole, as well as each item in that checklist, is complete.

Problem Statement

A number of accidents have occurred because of the failure of the flight crew to complete all items on a checklist. Many paper checklists even include "checklist complete" as their last item, to ensure that the checklist is complete. Ensuring that a checklist has been completed is an important function that an ECL should support as well. This includes indicating both the status of the checklist as a whole and each item on the checklist. If a checklist-level indicator is not provided, the crew must take the time to scan each item in the checklist. If an individual-item indicator is not provided, the crew cannot determine which items are incomplete if no "checklist complete" message occurs.

Example(s)

Checklist completion could be indicated in two ways:
- Visual indications that show all individual items are completed
- Visual indication that the checklist as a whole has been completed

Both completion indicators should be provided. An aural indicator may also be used.

Evaluation Questions
- Is the completion of all individual items indicated?
- Is the completion of the entire checklist indicated?
- If provided, is the aural indicator of checklist completion clear ?

4.4 Optional Features

4.4.1 Links Between Checklist Items and Related Information

Equipment Recommendation(s)

> A set of links to information related to individual checklist items should be provided. The links could direct users to additional information about that item, about the system addressed by the item, and/or to MEL information for that system. Returning to the checklist item from the related information should be a single-step action.

Equipment Issue(s)

> In choosing the type of information that is linked to a particular checklist item, designers should consider what type of information is most likely to be needed by the crew. This information should be easy to access. If the user is allowed to select the type of information to access, there should be a consistent method for making this selection.

Problem Statement

Even though crews are highly trained and practiced on the use of checklists, they may occasionally find that they need to look up related information. The ECL can make it much easier to find related information. This feature will increase usability as well as user acceptance of ECLs.

The related information could be about that item, the system addressed by the item, and/or the MEL information for that system. Additional information about a particular item may be especially useful for less experienced users. Links to related information about a particular system could be useful particularly as a study and review tool. Links to MEL information would provide immediate access to the implications of a non-normal condition identified through checklist performance.

Example(s)

A direct link to related information could be implemented by means of a hyperlink initiated from the checklist item. Another way to access related information would be through a pop-up menu that appears over the item when called up. The user could choose the information they wanted to view from this menu.

Evaluation Questions

- Does the ECL provide links to related information? Can the user select the type of information to view easily?
- Is the most useful information easy to access?
- Can the user return to the checklist from related information in one step?

4.4.2 Links to Special Information for Ongoing Non-Normal Conditions

Equipment Good Practice(s)

> Users should easily be able to access procedural, system, and operational notes and other information pertaining to any ongoing non-normal condition.

Problem Statement

Procedural changes are often made in response to a non-normal condition. In addition, some non-normal conditions can result in the loss of other systems. For example, in some aircraft, a hydraulic system 1 failure also causes the loss of the outboard ground spoilers. It can be challenging to remember this information because it may apply to subsequent phases of flight, and because crews are very busy in these situations, particularly when multiple non-normal conditions have occurred.

In order to relieve pilot workload and minimize the opportunity for error, ECLs could provide easy access to all procedural and operational notes and other appropriate information pertaining to an ongoing non-normal condition. This information could then be reviewed at the appropriate time, enabling the crew to prepare for each flight phase in advance. Customization of the information may be useful as well.

Example(s)

Ongoing non-normal conditions, inoperative systems, procedural changes, operational limitations, and other information could be listed in one area. Advanced versions of this function could allow the user to select which checklists to pull this information from or to choose specific paragraphs from operating manuals and other documents for inclusion.

Evaluation Questions
- Can the user access checklist information for non-normal conditions easily at any time?
- Can the user customize the information for a non-normal condition?

4.4.3 Links to Calculated Values

Equipment Requirement(s)

> If the EFB provides calculation worksheets, easy access from the checklist item to the corresponding worksheet must be provided to support initial calculation, as well as subsequent review and modification of the calculated value.
>
> ECL values that were calculated in a linked worksheet must appear in the corresponding checklist location. The corresponding checklist fields must be blank prior to insertion of the calculated value.

Training/Procedure Requirement(s)

> Procedures must be in place that define the roles the flight crew and dispatch play in creating and reviewing performance calculations supported by the ECL.

Problem Statement

Some checklist items involve setting a system value that has been calculated from one or more tables. The calculation worksheet may be immediately accessible from the corresponding checklist item. If the EFB supports performance calculations, the calculated value must appear in all appropriate checklist locations.

Example(s)

Direct access to the appropriate worksheet must be provided for all checklist items that can be calculated using the EFB. In addition, the user must be able to easily return to the checklist item from which the worksheet was accessed, even if the calculation was not attempted or completed.

Any checklist field that requires a calculation must be empty until the calculation has been completed. Once complete, the value must appear in all appropriate checklist locations. Access to the calculation worksheet must be supported even after the calculation has been completed to enable the user to review the assumptions on which the calculation is based.

In addition, the checklist item and the corresponding calculation worksheet should support access to background information that is useful to understand the calculation.

Evaluation Questions

- If the EFB provides worksheets, how are the appropriate worksheets accessed from the checklist items that require them, and is this easy to accomplish?
- If the EFB provides worksheets, how does the user return to the checklist item from the worksheet and is this easy to perform?
- If the EFB supports worksheets that are linked to specific checklist locations, is the checklist item
    - Empty prior to the performance of the calculation?
    - Filled in after the performance of the calculation?
- How is access to the worksheet supported if the user chooses to return to the calculation after it has been completed and is this easy to accomplish?
- Does the ECL provide background information that can help the user understand how to perform the calculation?
- Does the air carrier have procedures in place that define flight crew and dispatch roles in creating and reviewing performance calculations supported by the ECL?

4.4.4 Task Reminders

Equipment Recommendation(s)

> ECLs should provide reminders for tasks that require a delayed action to ensure that the task is completed at the appropriate time.

Equipment Requirement(s)

> If the ECL supports task reminders, the reminder must be displayed constantly once in progress, and it must attract the pilot's attention at the time that the delayed action should be performed.
>
> If multiple task reminders can be in progress at one time, crews must be able to determine how many are in progress and to what tasks they refer.

Problem Statement

Pilots may become distracted and forget to complete tasks that require a delayed action (e.g., *stopping* a fuel transfer). ECLs that provide a reminder to complete the delayed action can ensure that the task is completed at the correct time.

Example(s)

An example of a task that requires time to complete is a fuel transfer. After the fuel transfer is initiated, the crew typically completes other tasks while the transfer progresses, since the transfer may last several minutes. After the correct amount of time has passed, the crew must stop the fuel transfer, but by this time they may be engaged in other tasks. An ECL reminder that attracts their attention at the correct time could ensure that the fuel transfer does not go on too long.

The reminder could be a visual icon in one corner of the display. It could begin to flash to attract attention. An aural warning could also draw the crew's attention to the visual indicator.

Multiple task reminders could be indicated in several ways. One approach is to use an icon with a number on it that indicates the number of active reminders. The user could click on the icon to access a master list of all active reminders together with pertinent information such as expiration time. Another approach is to use one icon for each reminder. Clicking on the reminder would access more information about that task.

Evaluation Questions

- Does the EFB provide reminders for tasks that require a delayed action?
- If task reminders are provided, how are multiple reminders indicated and is it easy to determine what tasks they are associated with?
- Can the user easily review what a reminder is for?

4.4.5 Checklist Branching

Equipment Requirement(s)

> When a checklist branches based on a key decision, the selected branch must be clearly indicated.
>
> The user must be able to backup to the decision step and choose another decision branch to allow recovery from an erroneous choice.

Equipment Recommendation(s)

> Items that are not on the selected branch should not be selectable.

Problem Statement

While performing a complex checklist, the pilot may have to make key decisions at several points, and choose a branch of that checklist based on each key decision. Keeping track of the active items along a given branch mentally can be cumbersome and error prone.

ECLs that clearly highlight the selected branch of a checklist can ease the mental burden of keeping track of which items to perform. Also, by clearly encoding the selected branch, the pilot may be more aware of the decision he/she made to select that branch. If items that are not along the selected branch are deactivated, then the pilot cannot mistakenly perform them.

Example(s)

Decision branching can be indicated by means of a yes/no indicator in response to an explicit, clearly written question. Double negatives should not be used in the question. Based on the user's choice, the user is taken to the next set of items required for that situation. The user must be allowed to change his/her choice by backing up to the decision step in the checklist.

Items that are not along the selected branch could be encoded by a text color, such as a dim gray, or they could be hidden. If they are hidden, the user should be able to view these items for review on request.

Evaluation Questions
- Are decision branches clearly indicated within a checklist?
- Are the checklist questions written clearly?
- Can the user recover from choosing the wrong branch easily?

5 Flight Performance Calculations

5.1.1   Aircraft Performance Documentation

Training/Procedures Requirement

> Procedures must be developed to ensure that any information required to be available outside of the aircraft is transferred from the EFB at the appropriate time and place.

Equipment Good Practice

> Even if there is no legal requirement to be able to transfer data off of the EFB, the ability to do so is desirable.

Problem Statement

EFBs may be used to compute flight performance information. In some cases (e.g., Part 121 Operations), copies of this computed information must be deposited at a ground station prior to takeoff. Therefore, there must be a procedure for transferring the data from the EFB to some other media (e.g., paper), or onto a different ground computer.

Part 91 operators may not be legally required to transfer data off the EFB, but would benefit from the ability to do so.

Example(s)

Weight and balance information must be available in either paper or electronic format to airline personnel other than the crew. One way of handling this requirement is to send the information electronically from the EFB to company flight dispatchers. An alternative would be for the EFB to print out the required information onto paper, which would then be left at the point of departure.

Evaluation Question(s)

- What is the procedure for ensuring that, if necessary, EFB data can be stored outside of the device?

5.1.2 Data-entry Screening and Error Messages

Equipment Requirement(s)

> If user-entered data is not of the correct format or type needed by the application, these data must be discarded, and a user-oriented error message must be provided.
>
> The error message must convey which entry is suspect and specify what type of data is expected.

Problem Statement

While pilots who use an EFB may receive some training in use of the device, designers of the system should not expect users to be experts. Well designed error messages help to reduce the training time, promote acceptance of the device, and aid in recovery from errors. Not all entry errors can be caught with data screening, but such screening can be quite effective nonetheless. When an incorrect item is identified and cleared, only that piece of data should be discarded, not the whole set of entries related to the particular task in progress.

Example(s)

In entering flap settings for a takeoff performance worksheet, values outside the normal range of flap settings (e.g., a three digit, or alphabetic entry) should not be accepted. If an invalid entry is made, the error message could state that a number between 0 and 30 (or whatever the maximum flap setting is for that aircraft) is required.

Evaluation Question(s)

- Are errors in data entry identified clearly to the user?
- Does the error message clarify the type and range of data expected?

5.1.3 Support Information for Data Entry

Equipment Requirement(s)

> The units of each variable used in the software must be clearly labeled.
>
> The labels, formats, and units of variables used in the software must match the labels, formats, and units of the data available to the user.

Equipment Recommendation(s)

> If the user may need to refer to other computations or data stored within the application, that information should be in view, or else easily accessible.
>
> Definitions of specific terms used in the software should also be easily accessible (e.g., via hyperlinks, or pop-up information areas.)

Equipment Issue

> When data entry is required, the EFB designer should consider whether that data will be readily available to the user.

Problem Statement

The user should not have to manipulate what information is in view in order to complete a single logical task. All the information necessary to complete that task should be in view, or else easily accessible. If some necessary information is out of view, the user may make errors, or may not catch errors made previously.

System designers should consider whether the data requested will be readily available to the user, and whether the format available to the user matches the format expected by the software. Other support information includes the definition of terms (e.g., what is a "damp" runway?), and unit labels.

Example(s)

In computing various flight plan variables, the user may need to refer to route information, which should be in view or readily accessible.

Different airline operators may use different terminology to refer to variables used in various computations, such as weight and balance. The terminology used in the software should match that used by the airline in any other paperwork (e.g., maintenance, or fuel records).

Evaluation Question(s)

- Is all the information necessary for a given task presented together, or at least easily accessible while performing that task?
- Does the terminology used in the software match the terminology used in other operator documents?
- Are units clearly labeled?

5.1.4 When and How to Do Performance Calculations

Training/Procedures Recommendation(s)

> In order to reduce the possibility of errors, users should learn to complete performance calculations at particular times within the flight. They should also plan to complete the calculations in a certain order, and through a certain set of ordered steps.

Training/Procedures and Equipment Recommendation(s)

> In designing the flow of steps for completing a task in the software, designers should consider whether that flow is logical to the user and can be trained easily. Similarly, training on how to do a performance calculation should be matched with the steps in the software.

Equipment Good Practice(s)

> The software could prompt or remind the user of the order of steps for completing a task.

Problem Statement

Performance calculations are usually completed in a series of steps. In some cases, the steps must be completed in a particular order, but in other cases, the order of steps is not critical. When the order of steps is important, the software should prompt the user for information in that order. Even if the order of steps is not critical, users should be trained on the process for completing the task so that they are less likely to forget individual steps.

Users should also be trained on when to initiate and complete performance calculations with respect to the flight timeline.

The software should be designed to match the users expectations with regard to the order of steps in the process, and the training should be designed to reinforce this procedure for completing a calculation.

Example(s)

Weight and balance computations are usually done at the gate while passengers are boarding. However, the weight and balance data may change significantly at the last minute, and so should be easily modifiable just prior to takeoff.

Evaluation Question(s)

- Are users trained on when and how to do performance calculations?

5.1.5 Default Values

Equipment Recommendation(s)

> In general, default values should be based on the most conservative parameters for that calculation.

Equipment Issue(s)

> EFBs that are integrated with other cockpit systems may be able to acquire default values from these systems. If however, the communication with the other system is lost, designers must have a backup plan for assigning default values.
>
> The software should ideally prompt users to review default values carefully.

Problem Statement

Default values can be useful in speeding a routine calculation task. However, default values should be selected carefully because they may not always be carefully reviewed by users. One way to mitigate the potential negative effects of default values is to always select the most conservative values for that situation.

Example(s)

In computing landing performance, the default value for runway conditions could be the one requiring the longest runway length.

Evaluation Question(s)

- Are the default values conservative?
- Are any defaults obtained from other cockpit systems? If yes, what is the backup plan for assigning these values if communication with the other system is lost?

Appendix A: Related Literature

Section 1: <u>FAA Documents on the EFB</u>

(DRAFT 19 July 2000) AC 20/120-EFB Operational Approval and Use of Portable Flight Data Equipment "The Electronic Flight Bag" (EFB).

(DRAFT September 1999) FAA Information Memo: Policy Guidance for Installation Approval of Electronic Flight Bag Equipment. (Air Transport Directorate)

(DRAFT October 1999) AC-20-FIS: Safety and Interoperability Requirements for Flight Information Services (FIS) Equipment.

FAA Memorandum (14 May 1998) (by Mike DeWalt): Approval of Avidyne Navigation Display and DO-178B.

AC 25-11 (16 July 1987) Transport Category Airplane Electronic Display Systems.

AC 91.21-1 (20 August 1993) Use of Portable Electronic Devices Aboard Aircraft

FAA Document 98-AIR-DATIS: Safety and Interoperability Requirements for Digital-Automatic Terminal Information Service (Digital-ATIS).

Section 2:

<u>General</u>

Society of Automotive Engineers (SAE) Aerospace Behavioral Engineering Technology (G-10) Subcommittee (August 2000) *DRAFT Aerospace Recommended Practice (ARP) on Human Factors Criteria for the Design of Multifunction Display for Civil Aircraft* (ARP 5364).

Federal Aviation Administration Technical Center (January 1996). *Human Factors Design Guide for Acquisition of Commercial-off-the-shelf Subsystems, Non-Developmental Items, and Developmental Systems—Final Report and Guide (Sections 7 and 8).* http://www.tc.faa.gov/act-500/hfl/products_index.htm

Society of Automotive Engineers (SAE) Aerospace Behavioral Engineering Technology (G-10) Subcommittee. (1996) Aerospace Recommended Practice (ARP) on *Human Engineering Recommendations for Data Link Systems* (ARP 4791).

<u>System Messages, Errors and Alerts</u>

FAA Advisory Circular (16 August 1999) *Guidelines for Design Approval of Aircraft Data Communications Systems* (AC 20-140).

Federal Aviation Regulations (FAR 23.1322, FAR 25.1322, FAR 27.1322, FAR 29.1322)

Terrain Awareness and Warning System (TAWS) Technical Standard Order TSO C151a http://av-info.faa.gov/tso/

<u>Graphical User Interface Style Guides</u>

Microsoft Corporation. *The Windows™ Interface Guidelines for Software Design.* (1995). Redmond, Washington: Microsoft Press.

Apple Computer, Inc. *Macintosh™ Human Interface Guidelines.* (1992). Reading, Massachusetts: Addison-Wesley Publishing Company.

Open Software Foundation (1993). OSF/MOTIF™ Style Guide (Revision 1.2), Englewood Cliffs, NJ: Prentice Hall.

Section 3: <u>Electronic Documents</u>

NASA/FAA Operating Documents Project Final Report.

Dillon, A. (1994) *Designing Usable Electronic Text.* Taylor & Francis: Bristol, PA.

Rosenfeld, L. & Morville, P. (1998) *Information Architecture for the World Wide Web.* O'Reilly & Associates: Sebastopol, CA.

McKinley, T. (1997) *From Paper to Web.* Adobe Press, Macmillan Computer Publishing: Indianapolis, IN, USA

Section 4: <u>Electronic Checklists</u>

Degani, A. & Weiner, E.L. (1990). *Human Factors of Flight-Deck Checklists: The Normal Checklist.* (NASA Contractor Report 177549). National Aeronautics and Space Administration Ames Research Center: Moffett Field, California.

Degani, A. (1992). *On the typography of flight-deck documentation.* (NASA Contractor Report 177605). Moffett Field, CA: NASA Ames Research Center.

Federal Aviation Administration. (4/24/96) *Operational Use and Modification of Electronic Checklists.* (AC-120-64).

Appendix A: Related Literature

Section 1: <u>FAA Documents on the EFB</u>

(DRAFT 19 July 2000) AC 20/120-EFB Operational Approval and Use of Portable Flight Data Equipment "The Electronic Flight Bag" (EFB).

(DRAFT September 1999) FAA Information Memo: Policy Guidance for Installation Approval of Electronic Flight Bag Equipment. (Air Transport Directorate)

(DRAFT October 1999) AC-20-FIS: Safety and Interoperability Requirements for Flight Information Services (FIS) Equipment.

FAA Memorandum (14 May 1998) (by Mike DeWalt): Approval of Avidyne Navigation Display and DO-178B.

AC 25-11 (16 July 1987) Transport Category Airplane Electronic Display Systems.

AC 91.21-1 (20 August 1993) Use of Portable Electronic Devices Aboard Aircraft

FAA Document 98-AIR-DATIS: Safety and Interoperability Requirements for Digital-Automatic Terminal Information Service (Digital-ATIS).

Section 2:

<u>General</u>

Society of Automotive Engineers (SAE) Aerospace Behavioral Engineering Technology (G-10) Subcommittee (August 2000) *DRAFT Aerospace Recommended Practice (ARP) on Human Factors Criteria for the Design of Multifunction Display for Civil Aircraft* (ARP 5364).

Federal Aviation Administration Technical Center (January 1996). *Human Factors Design Guide for Acquisition of Commercial-off-the-shelf Subsystems, Non-Developmental Items, and Developmental Systems—Final Report and Guide (Sections 7 and 8)*. http://www.tc.faa.gov/act-500/hfl/products_index.htm

Society of Automotive Engineers (SAE) Aerospace Behavioral Engineering Technology (G-10) Subcommittee. (1996) Aerospace Recommended Practice (ARP) on *Human Engineering Recommendations for Data Link Systems* (ARP 4791).

<u>System Messages, Errors and Alerts</u>

FAA Advisory Circular (16 August 1999) *Guidelines for Design Approval of Aircraft Data Communications Systems* (AC 20-140).

Federal Aviation Regulations (FAR 23.1322, FAR 25.1322, FAR 27.1322, FAR 29.1322)

Terrain Awareness and Warning System (TAWS) Technical Standard Order TSO C151a
http://av-info.faa.gov/tso/

<u>Graphical User Interface Style Guides</u>

Microsoft Corporation. *The Windows™ Interface Guidelines for Software Design.* (1995). Redmond, Washington: Microsoft Press.

Apple Computer, Inc. *Macintosh™ Human Interface Guidelines.* (1992). Reading, Massachusetts: Addison-Wesley Publishing Company.

Open Software Foundation (1993). OSF/MOTIF™ Style Guide (Revision 1.2), Englewood Cliffs, NJ: Prentice Hall.

Section 3: <u>Electronic Documents</u>

NASA/FAA Operating Documents Project Final Report.

Dillon, A. (1994) *Designing Usable Electronic Text*. Taylor & Francis: Bristol, PA.

Rosenfeld, L. & Morville, P. (1998) *Information Architecture for the World Wide Web.* O'Reilly & Associates: Sebastopol, CA.

McKinley, T. (1997) *From Paper to Web.* Adobe Press, Macmillan Computer Publishing: Indianapolis, IN, USA

Section 4: <u>Electronic Checklists</u>

    Degani, A. & Weiner, E.L. (1990). *Human Factors of Flight-Deck Checklists: The Normal Checklist.* (NASA Contractor Report 177549). National Aeronautics and Space Administration Ames Research Center: Moffett Field, California.

    Degani, A. (1992). *On the typography of flight-deck documentation.* (NASA Contractor Report 177605). Moffett Field, CA: NASA Ames Research Center.

    Federal Aviation Administration. (4/24/96) *Operational Use and Modification of Electronic Checklists.* (AC-120-64).

Appendix B: Acronyms

AMJ	Advisory Material Joint
AQP	Advanced Qualification Program
ECAM	Electronic Centralised Aircraft Monitor
ECL	Electronic Checklist
EICAS	Engine Indication and Crew Alerting System
EFB	Electronic Flight Bag
FAA	Federal Aviation Administration
FMC/FMS	Flight Management Computer/Flight Management System
GUI	Graphical User Interface
JAA	Joint Aviation Authorities
LOE	Line Oriented Evaluation
MEL	Minimum Equipment List
PDF	Portable Document Format, a page-definition language
POH	Pilot Operating Handbook
PAI	Principal Avionics Inspector
POI	Principal Operations Inspector
QRH	Quick Reference Handbook

www.ingramcontent.com/pod-product-compliance
Lightning Source LLC
Chambersburg PA
CBHW081830170526
45167CB00007B/2766